# Lecture Notes in Artificial Intelligence    1456

Subseries of Lecture Notes in Computer Science
Edited by J. G. Carbonell and J. Siekmann

## Lecture Notes in Computer Science

Edited by G. Goos, J. Hartmanis and J. van Leeuwen

Springer
*Berlin*
*Heidelberg*
*New York*
*Barcelona*
*Budapest*
*Hong Kong*
*London*
*Milan*
*Paris*
*Singapore*
*Tokyo*

Alexis Drogoul   Milind Tambe
Toshio Fukuda  (Eds.)

# Collective Robotics

First International Workshop, CRW'98
Paris, France, July 4-5, 1998
Proceeedings

 Springer

Series Editors

Jaime G. Carbonell, Carnegie Mellon University, Pittsburgh, PA, USA
Jörg Siekmann, University of Saarland, Saarbrücken, Germany


Volume Editors

Alexis Drogoul
University of Paris 6
4 Place Jussieu, F-75232 Paris Cedex 05, France
E-mail: Alexis.Drogoul@lip6.fr

Milind Tambe
University of Southern California
Marina del Rey, CA 90292, USA
E-mail: tambe@isi.edu

Toshio Fukuda
Nagoya University
Furo-cho, Chikusa-ku, Nagoya, 464-01, Japan
E-mail: fukuda@mein.nagoya-u.ac.jp

Cataloging-in-Publication Data applied for


Die Deutsche Bibliothek - CIP-Einheitsaufnahme


**Collective robotics** : first international workshop ; proceedings / CRW '98,
Paris, France, July 4 - 5, 1998. Alexis Drogoul ... (ed.). - Berlin ;
Heidelberg ; New York ; Barcelona ; Budapest ; Hong Kong ; London ; Milan ;
Paris ; Singapore ; Tokyo : Springer, 1998
  (Lecture notes in computer science ; Vol. 1456 : Lecture notes in artificial
  intelligence)
  ISBN 3-540-64768-6


CR Subject Classification (1991): I.2.9, I.2.11, I.2

ISBN 3-540-64768-6 Springer-Verlag Berlin Heidelberg New York

© Springer-Verlag Berlin Heidelberg 1998
Printed in Germany

Typesetting: Camera ready by author
SPIN 10638245        06/3142 – 5 4 3 2 1 0     Printed on acid-free paper

# Preface

The collective robotics workshop (CRW'98) is an attempt to foster Distributed AI (DAI) and intelligent robotics research by examining a wide range of technologies devoted to Collective Robotics. Special attention has been focused on RoboCup, the widely acclaimed international effort for robotic and simulation soccer. CRW'98 is a key component of "Agents' World", which brings together several workshops and conferences emphasizing agent-based and multi-agent systems in numerous and diverse environments.

The main purpose of CRW'98 is to emphasize the relationships between DAI, ALife, and Collective Robotics. Following a rigorous review process, thirteen papers, focusing on such relationships, were selected for presentation at the workshop. These papers represent a truly international participation at the workshop, as well as diversity in the range of topics covered, as indicated in the call for papers:

* multi-agent techniques of cooperation for collective robotics
* multi-agent approaches to robots' control
* learning techniques for collective robotics
* methodologies for designing teams of robots
* planning for collective robotics
* self-organization, biologically inspired organizations for robotics systems
* ALife related researches for collective robotics
* multi-agent simulation environments for collective robotics
* simulated collective robotics (including simulated soccer playing robots)
* applications of collective robotics in industrial, military, and public domains
* models of behavior for autonomous robots
* cooperation, interaction between humans and robots
* micro- and nano-robotics
* collective robotics research's contribution to DAI and ALife
* robotic soccer as a standard problem for DAI and Robotics
* presentation of actual robotic soccer teams

Many people were involved in the organization of CRW'98. We would like to thank all of them, particularly all the members of the program committee, whose time and expertise enabled the selection of this fine sampling of research in collective robotics.

June 1998

Alexis Drogoul
Milind Tambe
Toshio Fukuda

# Table of Contents

# Exhaustive Geographic Search with Mobile Robots Along Space-Filling Curves[1]

Shannon V. Spires
Steven Y. Goldsmith

Advanced Information Systems Laboratory
Sandia National Laboratories
Albuquerque, New Mexico USA
*svspire@sandia.gov, sygolds@sandia.gov*

**Abstract.** Swarms of mobile robots can be tasked with searching a geographic region for targets of interest, such as buried land mines. We assume that the individual robots are equipped with sensors tuned to the targets of interest, that these sensors have limited range, and that the robots can communicate with one another to enable cooperation. How can a swarm of cooperating sensate robots efficiently search a given geographic region for targets in the absence of *a priori* information about the targets' locations? Many of the "obvious" approaches are inefficient or lack robustness. One efficient approach is to have the robots traverse a space-filling curve. For many geographic search applications, this method is energy-frugal, highly robust, and provides guaranteed coverage in a finite time that decreases as the reciprocal of the number of robots sharing the search task. Furthermore, control is inherently decentralized and requires very little robot-to-robot communication for the robots to organize their movements. This report presents some preliminary results from applying the Hilbert space-filling curve to geographic search by mobile robots.

## Introduction

The idea of using swarms of cooperating robots to solve various physical problems has received much attention recently [CFKM95], and has recently become cost-effective and practical with the advent of less expensive hardware and with new analysis techniques from complexity theory, chaos theory, and nonlinear dynamics. One problem that is particularly applicable to robot swarms is exhaustive geographic search. Exhaustive geographic search asks that we develop a complete map of all phenomena of interest within a defined geographic area, subject to the usual engineering constraints of efficiency, robustness, and accuracy [GR98b], [CP97], [HTL96], [KOAY95].

One example of such a search problem is that of finding buried land mines. It is possible to build a robot that has the requisite sensors and navigation apparatus to do this job, thus removing humans from an extremely high-risk activity. But a single robot is likely to be expensive and subject to damage from the mines, rendering it useless. If less expensive robots could be built in quantity, it would be better to build a swarm of such robots that could cooperate to locate all the mines within a given geographic region. With the right programming, many robots could do the job faster than one (ideally in $1/m$ the time, if $m$ robots participate). The search process would

---

[1] Sandia is a multiprogram laboratory operated by Sandia Corporation, a Lockheed Martin Company, for the United States Department of Energy under Contract DE-AC04-94AL85000.

be more robust because if a robot is damaged by a mine, others could take over its work.

This report presents an efficient, robust, cooperative search algorithm that allows multiple robots to be applied to such a mission. The primary insight is that exhaustive coverage of a one-dimensional path with multiple robots is substantially easier and more robust than coverage in two (or more) dimensions. This algorithm transforms a two-dimensional search task into one-dimensional search, with the result that two dimensions can be effectively searched with the attendant advantages of one-dimensional coverage. The transformation is accomplished by having the robots traverse a minimum-length space-filling curve, dividing the curve among $m$ robots.

## Problem Constraints

The nature of the problem is such that there is no *a priori* knowledge of the locations of the targets (the targets are the land mines in the example above); we merely know that there *may* be one or more targets within a given geographic region. Our mission is to discover definitely their total number, their locations, and optionally to further characterize them along other dimensions. The requirements and constraints of the task are as follows:

- Each robot has a sensor apparatus which is adequate for detecting the targets of interest, but its range is limited to an area much smaller than the overall region of interest. We assume that all the robots have the same sensor range, or that they can all be made aware of the minimum sensor range of any of the robots.
- Each robot can reliably communicate with the other robots with some probability $P_c$. The search mission should be completable even as $P_c$ approaches 0, although it may take longer.
- Each robot has a finite energy supply (e.g. batteries or a fuel cell).
- The robots "know" when their job is finished.
- The robots do not collide.
- The robots stay within the bounds of the search region.
- A robot's individual search area should not overlap[2] that of another robot, but the entire search area must be covered (hence the use of the term *exhaustive* search).
- The search should still be completable even if one or more robots is damaged or otherwise fails during the mission. As long as at least one robot remains alive, the search should complete, although it may take longer.
- We must accomplish the search within a finite time.
- The search time should decrease as the number of robots participating increases.
- Control of the robots is completely decentralized; there is no central coordinating entity. The robots must be able to behave autonomously. Isolated robots must still be able to perform useful work.

## Geographic Search

In general, a geographic search will take place in three phases: (1) Initial

---

[2] There may be cases where, for reasons of higher sensor reliability, we explicitly *do* want them to overlap.

configuration; (2) Search; (3) Terminal configuration [GR98a]. First, the robots must organize themselves into a coherent, communicating group within the region. They divide the area into subregions and assign each robot one or more subregions. Next, the search task proper is conducted. Finally, the robots determine through consensus that the region has been searched. At this point, the robots take some terminal action.

To maximize efficiency, two considerations are immediately obvious:

A) *Minimal configuration energy.* We don't want the robots to expend a great deal of energy in the initial configuration phase because no useful work is accomplished until the search phase. During initial configuration, the robots must expend energy communicating with each other and moving into their initial positions. These energy expenditures must be minimized.

B) *Optimal search coverage.* During the search phase, we must ensure that the entire search area is covered in a finite time, preferably proportional to the reciprocal of the number of robots involved.

To optimize (A), we could have the robots perform no initial configuration at all, with each simply searching the area in its immediate vicinity. No initial communication and no initial movement take place to coordinate with the other robots. We assume the robots are initially randomly distributed on the search region (because they were dropped out of an airplane, for example). Each robot could perform a random walk [W97a], spiral outward, or use some other autonomous algorithm. This defeats the purpose of multi-robot collaboration. Since there is no coordination, areas already searched can be retraced. Spiraling outward would guarantee that the search area was covered in a finite time, but using more than one robot would not necessarily decrease that time. A random walk is even worse—it wouldn't even guarantee coverage in a finite time.

Optimizing (B) means having the robots subdivide the search area equally and each search its agreed-upon subregion. A good deal of initial collaboration and movement would generally be needed here, especially if the robots start in random locations and must move to an organized initial configuration. An example of such an initial configuration is shown in Figure 1, where eight robots (triangles) have lined up in a column along the left side of the search region and plan to march to the right. Each search subregion is a horizontal strip.

**Figure 1:** Robots (triangles) aligned in preparation to march to the right

This accomplishes a maximally-efficient search, but at the cost of a less efficient initial configuration. In addition, this search assumes all the robots are highly reliable; it's not very robust if one or more robots dies during the search. (We'll discuss why later in the report.) The dilemma is thus to find the optimal tradeoff between the initial configuration phase and the search phase that makes both acceptably efficient, while ensuring the robustness of the mission.

We believe that searching along a space-filling curve solves the dilemma nicely: it requires very little initial configuration while also providing an efficient, robust search. Other work in coverage [CP97, HTL96] addresses obstacle avoidance (currently a weak point of our algorithm) but does not address efficient coverage with multiple robots or robustness. [KOAY95] discusses efficient multiple-robot sweeping but control is not distributed and robustness is not addressed.

We will digress a bit here to describe space-filling curves in general.

## Space-Filling Curves

A space-filling curve [S94] is a one-dimensional curve which passes through every point of a given N-dimensional region. It is thus a one-to-one mapping between an N-dimensional and a 1-dimensional space. Such curves can be constructed to fill regions of any dimensionality, but in our case we're mainly interested in two-dimensional regions, corresponding to a geographic area on the surface of the earth. Two-dimensional regions are also much easier to illustrate, and for those reasons the remainder of this report will concern only curves that fill 2-d regions. The reader should bear in mind, however, that the constructions and algorithms described herein can easily be extended to regions of more dimensions.[1]

Of course, a true space-filling curve is an ideal mathematical entity and can only be approximated in the real world. If we divide our region of interest into subregions of finite size, rather than points, we can easily draw a curve which passes through each subregion. In Figure 2, for example, a square region is divided into four subregions. The dark inverted U-shaped curve passes through the center of each subregion.

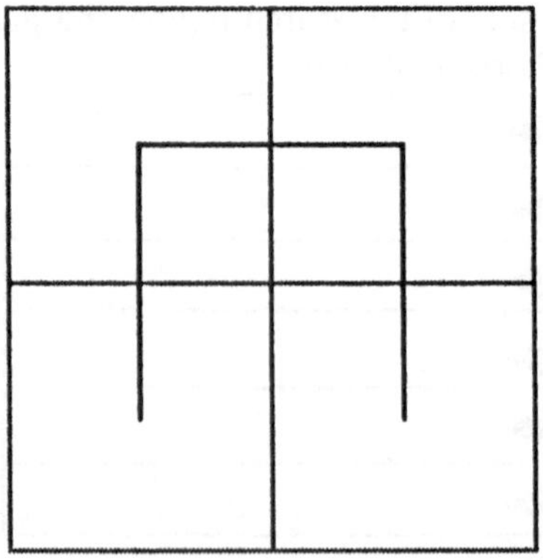

**Fig. 2.** First Order Hilbert curve

---

[1] A 3-d region [G97], [SLP83] might be appropriate for search by a swarm of undersea [HTL96] or flying robots, for example.

Subdividing further, into 16 regions (4 x 4) the curve becomes as in Figure 3.

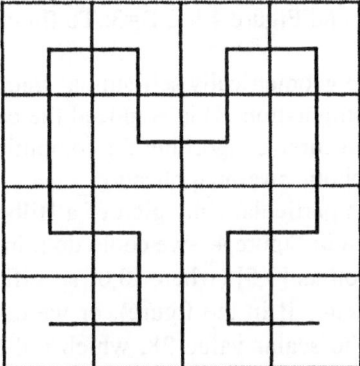

**Fig. 3.** Order 2 Hilbert curve

Subdividing still further, into an 8 x 8 grid, Figure 4 shows a curve which passes through each of the 64 subregions:

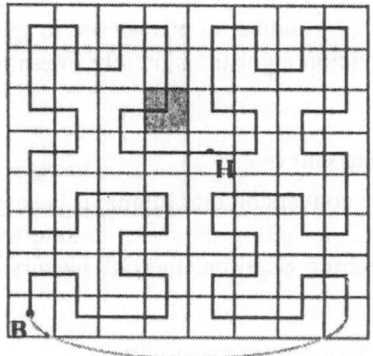

**Fig. 4.** Order 3 Hilbert curve (gray arc at bottom is explained in the text)

The family of curves illustrated in Figs 2-4 is called the *Hilbert* [W97b] curve after David Hilbert, their discoverer. There are many families of space-filling curves, but the Hilbert has some especially nice properties. For example, two points that are "close" in 2-d space are more likely to be "close" in Hilbert space than with other space-filling curves [MJFS96].

Notice how every deeper subdivision of the curve contains four copies of the entire previous curve, suitably rotated and reflected, with some straight segments added to ensure continuity. The curve is thus geometrically self-similar, as space filling curves frequently are. The *order* of the curve $D$ determines the number of subregions it passes through and consequently, the total length of the curve:

$$\# \text{ subregions} = (2^D)^2 = 2^{2D} = \text{Length}_{\text{curve}} + 1$$

Thus Figure 2 is a first order curve, or D=1, because it occupies 4 subregions and is 3 units long. Figure 3 has D=2, and Figure 4 has D=3. To fill every point in a region, a curve would have to be D=∞.

The fact that the curve is a geometrically self-similar fractal implies that there is a recursive algorithm for its computation. This is indeed the case [McW97], [PLF91]. However, there is also a non-recursive algorithm for computing a given Hilbert curve which is more useful in our robotic swarm application.

If we wanted to identify a particular subregion of a Hilbert-traversed space—for example, the shaded subregion in Figure 4—we could do it in two different ways. We could specify its [x,y] location as [3,5] (where [0,0] specifies the lower left corner subregion marked with the letter B in the figure), or we could specify its position along the Hilbert curve as the scalar value 28, which indicates that it is the 28th subregion encountered along the curve measuring from the beginning of the curve at subregion B. Subregion B is location 0 in this frame of reference.

There is a straightforward algorithm to convert the 2-space coordinates of real space into a 1-space Hilbert coordinate and an equally simple algorithm to convert from 1-space back into 2-space [B69]. Even though the dimensionality changes during this conversion, no information is lost because the total number of possible bits in x or y is D, while the total number of bits needed to specify the 1-space Hilbert coordinate is 2D. The conversion algorithms merely interpret the coordinate bits in different ways.

## Search Along a Space-Filling Curve

By using 1-space to 2-space conversion algorithms, it is easy to program a robot to follow a space-filling curve. Assuming the robot is able to deduce its current x,y location (via Global Positioning Satellite, inertial sensors, etc.), the algorithm is shown in Table 1:

$x_0,y_0 \leftarrow$ current location
$H \leftarrow$ Convert-xy-to-Hilbert($x_0,y_0$)
$H \leftarrow H + 1$
$x_1,y_1 \leftarrow$ Convert-Hilbert-to-xy(H)

**Table 1:** Hilbert traversal algorithm. $x_1,y_1$ represents the robot's new location.

In practice, we perform the operation $H \leftarrow H + 1$ modulo $2^{2D}$, which causes the robot to return to the beginning point of the curve after it reaches the end; this closes the curve and makes it a topological circle instead of a line segment. The gray curved arc at the bottom of Figure 4 reflects this closing.

If we now give the robot a sensing mechanism whose sensory range is at least as large as a single subregion, by traversing the curve it can now search the overall region in time proportional to the number of subregions, $2^{2D}$.

Notice that this algorithm works regardless of where on the curve the robot starts; it need not start at the H=0 point of the curve. This is one reason why the recursive algorithm is not used in the robotic application; it would be difficult to initialize the recursive control stack properly to allow traversal to begin at an arbitrary point.

Starting at an arbitrary point becomes valuable when more than one robot is involved. If we have two robots, we could start one at the beginning of the curve (point B in Figure 4) and the other at the halfway point of the curve (point H in Figure 4); the resulting traversal of the curve takes place in half the time it would take with one robot. In general, if we start with $m$ robots, we can search the space in time proportional to $2^{2D}/m$, provided we initially space the robots equally along the curve.

**Efficiency**

We'll now examine how search along a space-filling curve meets efficiency criterion (A). If we assume an initial random configuration of the robots within the region, we can actually get away with almost no initial configuration movement whatsoever. We also assume that the search region is square[4] and that its bounds and the robots' individual sensor ranges are known by each robot *a priori*.

Configuration proceeds as follows:

1) Each robot senses its own initial location, either absolutely or relative to the search region. This is a fundamental assumption of the algorithm and should be easy to realize with Global Positioning Satellite receivers or some other mechanism.

2) Each robot independently computes the subregion size based on its sensor range.[5] Since each also knows the bounds of the search region, each also computes its own location relative to the search region (unless its location sensor already provided the relative location) and computes the order of the Hilbert curve based on the size of the subregions relative to the overall search region. Again, each robot will compute the same value for the order of the curve.

3) Each robot decides which subregion it is currently within by quantizing its x,y location relative to the overall region. It then moves to the center of that subregion. This is the only time its wheels need to turn during the initial configuration phase, and the maximum distance any robot will have to move is

$$\frac{s}{\sqrt{2}}$$

where $s$ is the size of a subregion square measured along a side.

4) The robots broadcast their individual initial locations to each other so that each knows the starting locations of all the others. (How the robots reliably communicate this information is itself an interesting question but is beyond the scope of this report [G93]. The clustering properties of the Hilbert curve may be useful here in case the robots are too far apart to achieve total communication.)

Configuration is now complete and the actual search begins. Each robot proceeds to each subregion in turn along the Hilbert curve, using the algorithm in Table 1. At each subregion, the robot senses any targets of interest and checks to see if its next

---

[4] Non-square regions and even general polygonal regions can be handled by extensions of the techniques presented here.

[5] We assume for now that all the robots have the same sensor range, and thus compute the same subregion size.

subregion is the starting subregion of any other robot. If so, it stops and broadcasts "I'm Done!" along with information about any targets it may have found along the way. When all the robots have reported in, the collective can by consensus decide to report back to a higher authority, perform other tasks, or move on as a group to search another region. If after some reasonable time one or more robots have not reported in, the robot immediately behind the nonreporting robot can autonomously decide to proceed into the nonreporting robot's search area, search it, and report "I'm Done!". Note that each robot can precompute its own expected time to complete its portion of the search, since it knows at the outset how much of the curve it must traverse. These timeout values can be communicated to allow the collective to detect such fail-stop [K95] conditions.[6]

This leads us into an analysis of the expected search time. Certainly the time required to achieve guaranteed exhaustive coverage of the search region is bounded. At worst, the search will take time proportional to the number of subregions in the search region (or equivalently, the length of the space-filling curve). The precise upper bound time, $T_{UB}$, can be calculated by multiplying the expected movement speed of a robot by $s$ (the size of a subregion) by the total number of subregions. At best, the lower bound time will be $T_{LB} = T_{UB}/m$, with $m$ the number of robots. The expected exhaustive coverage time $T_E$ will equal $T_{LB}$ when the robots are initially configured to be spaced equidistant from each other along the space-filling curve. But we're not doing that; we're just moving each to the center of the subregion it initially occupies.[7] In this kind of initial configuration, $T_E > T_{LB}$ but is still nowhere near $T_{UB}$. Empirical results based on Monte Carlo simulation are shown in Figure 5. The graph clearly shows a reciprocal relationship between $T_E$ and number of robots. It shows that with 20 robots, the expected time to exhaust the space is less than 20% of the time required with 1 robot. This is obviously not as good as the theoretical best case of $T_{LB}$ = 5%, but it's not bad for almost no initial configuration movement. We believe that with *slightly* more initial configuration movement, $T_E$ can come much closer to $T_{LB}$. The nature of the particular space-filling curve being used becomes very important here. This is an open issue we are investigating.

---

[6] The issue of Byzantine failures [K95], [LSP82], in which a robot pretends to cooperate but actually is a saboteur, must also be addressed in environments where malevolent robots may be present.

[7] We assume for the sake of simplicity that no two robots initially occupy the same subregion. For a real implementation, a resolution protocol would be needed to deal with this possibility.

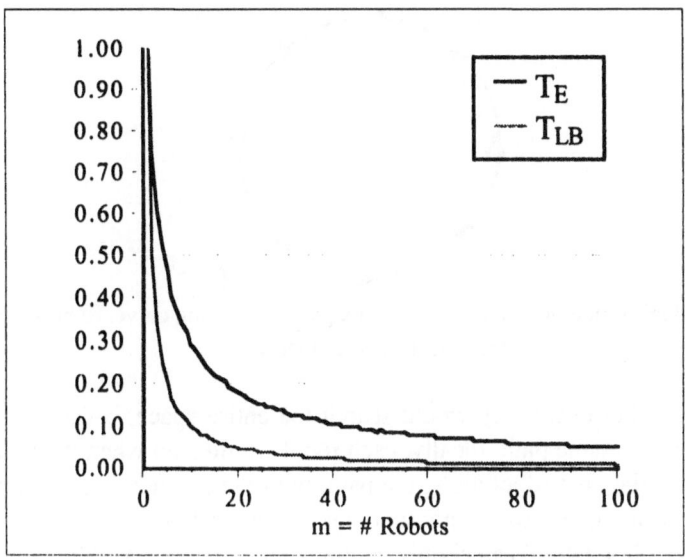

**Fig. 5.** $T_E$ (upper curve) as a percent of $T_{UB}$ vs. m. ($T_{LB}$ is shown for comparison purposes.) Random initial positions and minimal initial configuration movement. Empirical results.

Another point is that with the Hilbert curve, the search region must be divided into an integer power of 2 subregions (measured along a side). This can require the subregions to be significantly smaller than the sensor radius, thus increasing the overall distance traveled beyond that which would be required in the ideal case. Other curves which are not recursively subdivided by 2 but by other factors could remove this inefficiency, as could multi-level decomposition of the search region into macro- and micro- regions, where each type of region has a different coverage algorithm.

**Robustness**

In the initial configuration shown in Figure 1, where a space-filling curve is not being used, what happens if one of the robots is defective, or dies during the march? Its strip of territory will not be searched. The other robots will have to reliably detect this and reorganize themselves to search the defective robot's area. But if the robots are searching along a space-filling curve, all the robots are topologically searching in one dimension along a single circle (Figure 6).

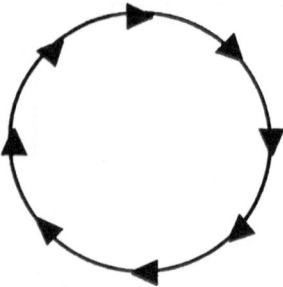

**Fig. 6.** Topological interpretation of search along a space-filling curve. Robots (triangles) traverse the curve clockwise.

Here, each robot eventually would search the entire space, because they are all searching along the same path. Ideally, each robot would stop when it encounters the starting point of the next robot along the path, with the result being that the space is searched in $1/m$ the time it would normally take with 1 robot. But if one robot breaks down, the next robot behind can simply continue along its path without stopping, and the overall space still gets searched, albeit in slightly longer time. No reconfiguration of the robots is necessary to accomplish this. Indeed, if as many as $m$-1 of the robots all break, the search is still guaranteed to complete (assuming the surviving robot has enough energy reserves to search the entire space).

The search is also robust with respect to communication breakdowns. If all the robots can communicate with each other, the only groupwide communications needed are for initial configuration, and when a robot encounters the starting point of the next robot along the path[8], it must announce to the collective "I'm Done!". When all robots have reported in, the collective knows the region has been completely searched. If one or more robots have not reported in by some time limit, the collective can deduce that they are dead and the next robot behind each dead robot can unilaterally decide (with no need for further communication with the group) to search the dead robot's area. If all communications fail (because of RF jamming, for example) the worst case is that the search will take as long as it would with only a single robot. Each robot will end up searching the entire space, and each will know to stop when it reaches *its own* starting point. The search will take a long time, but it's still guaranteed to complete.

Thus communications failures can cause the search mission to take longer, but they cannot prevent it from being accomplished.

## Collision Avoidance

In any collaborative robotics system, collision detection and avoidance will be needed. Space-filling curves may make collision avoidance easier. When several robots follow a single space-filling curve, there is theoretically no chance of two robots' paths crossing, and therefore no need for dynamic collision avoidance. Realistically, two robots could collide if navigation errors are allowed to accumulate,

---

[8] It can know where the next robot's starting point is either by memorizing the location during the initial configuration communication, or by having each robot drop a marker [R97] on the ground at its own starting point. In the latter case, initial communication may be unnecessary.

if two adjacent robots along the curve travel at different speeds, or if the one in front dies. But it's possible that collision detection could be simplified because of the automatic separation inherent in the algorithm.

**Further Research**

Much work still needs to be done exploring the general problem of robot traversal of space-filling curves. Some avenues of exploration include obstacle avoidance, formal expected exhaustion time analysis, handling robots with differing sensor ranges, non-square and non-rectangular search regions, and automatic determination of the search region based on energy stores and initial positions of the extrema robots in a cluster. Obstacle avoidance might be done by adapting the techniques described in [CP97], [HTL96], and [KOAY95]. Space-filling curves other than the Hilbert need to be investigated as well. One disadvantage of the Hilbert curve is the long straight path that connects its endpoints when it is closed. Other curves have the property that their beginning and end points are adjacent. Also, in cases where there is a penalty for turns, other space-filling curves may do the job with fewer turns than the traditional Hilbert. The Hilbert II curve [W97b] might be a candidate here.

And of course, the tradeoff between minimal initial configuration energy vs. minimal search energy must be further explored in light of particular applications.

**Conclusions**

We have begun to explore how space-filling curves can enhance the efficiency and robustness of geographic search by robot collectives. Initial results have been quite promising, especially in applications such as mine-clearing where exhaustive search is necessary. Much work still needs to be done, but it appears that combining this kind of search with more information-exploitive search algorithms [GR98b] may be extremely useful in general real-world search situations.

We hope to have movies of the robot simulations, as well as an expanded version of this report, online soon at http://www.sandia.gov/aisl/robotics/.

**References**

1. [B69] Bially, T. Space-Filling Curves: Their Generation and Their Application to Bandwidth Reduction. *IEEE Transactions on Information Theory,* V 15 No. 6, November 1969, pp. 658-664.

2. [CFKM95] Cao, U., Fukunaga, A., Kahng, A., and Meng, F. 1995. Cooperative mobile robotics: Antecedents and directions. *Proc. of IEEE/RSJ IROS,* pp. 226-234.

3. [CP97] Choset, H., and Pignon, P., Coverage Path Planning: The Boustrophedon Cellular Decomposition. International Conference on Field and Service Robotics, Canberra, Australia, 1997. Also available at http://voronoi.sbp.ri.cmu.edu/~choset

4. [G93] Gage, D., How to communicate to zillions of robots. *Mobile Robots VIII, SPIE,* 250-257, 1993.

5. [G97] Gilbert, W., *A Cube-Filling Hilbert Curve,*
   http://math.uwaterloo.ca/~wgilbert/Research/HilbertCurve/HilbertCurve.html

6. [GR98a] Goldsmith, S., and Robinett, R. *Collaborative Search by Mobile Robots, Part I: Problem Definition.* Sandia National Laboratories Technical Report. Preprint at http://www.sandia.gov/aisl/robotics/collaboration.

7. [GR98b] Goldsmith, S., and Robinett, R., *Collective Search by Mobile Robots Using Alpha-Beta Coordination.* Collective Robotics Workshop '98, Agent World, Paris, 1998.

8. [HTL96] Hert, S., Tiwari, S., and Lumelsky, V. A Terrain-Covering Algorithm for an AUV. *Autonomous Robots,* 3:91-119, 1996.

9. [K95] Kesteloot, L. *Fault-Tolerant Distributed Consensus.*
   http://tofu.alt.net/~lk/290.paper/290.paper.html

10. [KOAY95] Kurabayashi, D., Ota, J., Arai, T. and Yoshida, E. Cooperative Sweeping by Multiple Mobile Robots, *1996 IEEE Intl. Conf. on Robotics and Automation,* pp. 1744-1749, 1996.

11. [LSP82] Lamport, L., Shostak, R., and Pease, M. The Byzantine Generals Problem, *ACM Transactions on Programming Languages and Systems 4,* 3 (July 1982), 382--401.

12. [MJFS96] Moon, B., Jagadish, H. V., Faloutsos, C. and Saltz, J. *Analysis of the Clustering Properties of Hilbert Space-filling Curve,* 1996 University of Maryland Technical Report CS-TR-3611, http://www.cs.umd.edu/TR/UMCP-CSD:CS-TR-3611

13. [McW97] McWhorter, W., *Fractint L-SystemVariations,*
   http://spanky.triumf.ca/www/fractint/lsys/variations.html

14. [PLF91] Prusinkiewicz, P., Lindenmayer, A. and Fracchia, F. D., Synthesis of Space-filling Curves on the Square Grid. *Fractals in the Fundamental and Applied Sciences,* edited by Peitgen, H.-O. et al., Elsevier Science Publishers, 1991.

15. [R97] Russell, R. A., Heat Trails as Short-Lived Navigational Markers for Mobile Robots, *1997 IEEE Intl. Conf. on Robotics and Automation 1997,* pp. 3534-3539, 1997.

16. [SLP83] Stevens, R. J., Lehar, A. F. and Perston, F. H. Manipulation and presentation of multidimensional image data using the Peano scan. *IEEE Trans. on Pattern Analysis and Machine Intelligence,* PAMI-5(5) (1983) pp. 520-526.

17. [S94] Sagan, H. *Space-filling Curves,* Springer-Verlag, New York 1994.

18. [W97a] Weisstein, E., *Random Walk: 2-D,*
   http://www.astro.virginia.edu/~eww6n/math/RandomWalk:2-D.html

19. [W97b] Weisstein, E., *Hilbert Curve,*
   http://www.astro.virginia.edu/~eww6n/math/HilbertCurve.html

# A Multiagent System
# Based on Heterogeneous Robots

Andreas Birk and Tony Belpaeme

Vrije Universiteit Brussel
Artificial Intelligence Laboratory
Pleinlaan 2, 1050 Brussels, Belgium
{cyrano, tony}@arti.vub.ac.be
http://arti.vub.ac.be

**Abstract.** The paper presents the introduction of heterogeneity into a robotic Multiagent System. The system is based on ideas from Artificial Life; it forms a kind of artificial ecosystem where the animats are linked together in their search for "food" in form of electrical energy. Within this view, different robot-types can be seen as different species.
Given a basic robotic ecosystem with homogeneous agents — the so-called "moles" —, the two new species "mouse" and "head" are introduced. The differences between species are quite substantial. The "moles" for example have simple sensing capabilities whereas the "mouse" is equipped with vision. Some agents do not have social capabilities, whereas "heads" depend on cooperation. The paper describes how these differences and the common dependence on a global energy source interfere, and which conceptual and technological choices have to be made to keep a kind of ecological balance.

## 1 Introduction

We are interested in Multiagent Systems (MAS) from an Alife-perspective, i.e., MAS where behavior-oriented agents interact in ecosystem-like settings. In this view, the main goal of an agent is self-preservation, i.e., to stay operational [13, 19]. As resources, especially energy, are limited in time and space, agents have to compete for them. This forms the basis of all agent interactions in the system.

Most of the Alife research featuring ecosystems is based on simulations, e.g. [2, 4, 10, 11, 25, 27]. In this case, perception and effector-control of the agent are not embedded in the real-world. This causes various disadvantages which have been widely discussed before, and which are roughly summarized by the following quotation from Rodney Brooks: "the world is its own best model" [5]. Therefore, we follow the animat-approach [28] by using real robots.

Research on robotic ecosystems usually deals with homogeneous agents [6, 12, 17, 26]. We are interested in a heterogeneous MAS for several reasons. First, heterogeneity substantially adds to the complexity of the environment, which is of key interest to Alife-oriented research [7, 28]. Second, ecosystems with just one species are hardly biological plausible [14]. Lynne Parker also works on heterogeneous robotics agents. In doing so, she focusses on the aspects of a suited software architecure for this purpose (see e.g. [18]). We are, especially in this paper, more interested in the technical and conceptual problems of integrating new robotic agents into a concrete set-up. In doing so, we approach the subject more from an ALife than a software engineering perspective, and we focuss more on hardware issues.

We use an artificial ecosystem featuring animats from three different species. The differences are substantial as they include much more than just size and weight. First of all, the animats in this ecosystem have very different hardware-features as some have e.g. a vision module, others not. In addition, they differ in their behavioral complexity. Some can directly access a charging-station with simple reactive behaviors for re-filling their batteries, others depend on social interactions in the form of cooperation.

The rest of the paper is organized as follows. Section 2 describes the basic set-up of the VUB ecosystem including a charging-station, simple robots, and the so-called competitors. In section 3 the difficulties in the introduction of new species are presented. In section 4 the so-called "mouse" is described. This robot features vision which is used to detect the charging station and competitors. Section 5 introduces a somewhat unusual autonomous robot as it is immobile. This so-called "head" consists of a camera on a pan-tilt-unit with quite some vision capabilities. It sells information to other robots to prevent itself from starving. Section 6 concludes the paper.

## 2 The basic ecosystem

"Food" in form of electrical energy is the basic force merging the individual animats into our ecosystem. It is a crucial component to keep the animats *viable* [1, 13, 16]. Most of the animats are mobile[1] and powered by secondary batteries. These batteries can be re-filled in a charging station. Picture 1 shows the basic set-up as introduced and explained in detail in [15, 20]:

The basic ecosystem consist of

**the charging-station** where robots can autonomously re-fill their batteries. The charging-station is equipped with a bright white light on top.

**simple mobile robots** , the so-called "moles"[2] which are equipped with touch-based and active IR obstacle avoidance. In addition, they do phototaxis towards the charging station and towards

**the competitors** , these are boxes housing lamps connected to the same global energy source as the charging station. They can be dimmed by other inhab-

---

[1] One animat, the so-called "head", is not mobile and therefore has to rely on other robots. It is described later in this paper.

[2] The names for this and following species should not be taken too literal. The name "moles" attributes to the fact, that these robots have no vision. We use these names for reasons of convenience, not to imply or suggest any deeper relation with the biological versions.

**Fig. 1.** A part of the basic ecosystem with charging station, two robot "moles", three competitors, and several bricks as obstacles.

itants of the ecosystem by being pushed. The competitors establish a kind of working task which is paid in energy.

The basic ecosystem already provides various possibilities for Alife research. The interested reader is referred to e.g. [3, 21–23].

## 3 Heterogeneous agent-types

Heterogeneity is derived from the Greek words hetero (other) and genos (kind). So, the first question in our case is how kinds of robotic agents are classified. A taxonomy of agents can be based on the absence and presence of certain capabilities. Capabilities such as vision, learning, planning, social interactions, and so on. We denote the fact of having such a capability as *feature*. Roughly speaking, a feature is any kind of function that helps the agent to allocate resources in the environment. Within this view, the type of an agent can be defined by the sum of its features.

As mentioned before, self-preservation is a key issue for an agent in our system. At a first glance, features and self-preservation are linked in a straightforward way: the more features the better the agent is suited to survive as they help to allocate resources. But as we deal with robotic agents, features are grounded in the real world. It follows that not only features like "arm with gripper" are physical, even "mental" ones like learning can and have to be traced down to the physical level of e.g. the appropriate amount of computing power and memory needed for this task. So, additional features lead to an additional consumption of resources as well. It follows, that the agent-designer should not add features at liberty. Instead, agent-types have to match their environment and vice versa. Within the view of self-preservation, a feature makes only sense if its contribution to the allocation of resources is greater than its consumption of resources, i.e., its benefit is greater than its cost. Hardware factors are therefore not minor technical details, but crucial properties of an agent-type.

# 4 The feature vision and the "mouse"

Among the most crucial features for any animat are sensing-capabilities. Unfortunately, most of todays sensors are very specialized. Embedding adaptation on the sensor-level is somehow hard. Of course, many applications of learning to sensors are known. But they usually deal with increasing accuracy and robustness. Changing the functionality, like e.g. turning an active IR-sensor from an obstacle-avoiding device into an edge-detector, is infeasible for most sensors.

This does not hold in respect to vision. The same combination of a camera, a digitizer, and a computing-device can be used for obstacle-avoidance as well as edge-detection. Therefore, we chose vision as sensor-feature to play an important role in the "diversification-process" in the ecosystem.

The first new species equipped with vision is the so-called "mouse" (figure 2). This agent-type is a kind of enhancement of the "moles" by adding this feature based on a camera at the front of a robot. As explained in the previous section, we have to care about the trade-off between the benefit and the costs of vision in respect to resources, especially energy. Vision helps to allocate "food" as it can be used to perceive the charging station over larger distances and more accurately than by photo-taxis. But the hardware-costs (in terms of energy-consumption) have to be kept low to gain a real benefit. Imagine a basic robot, a radio-link, and an "invisible" number-crunching Cray next door doing elaborated vision-processing. Such a "dinosaur" would hardly be competitive to the "moles" as its feature vision has an immense additional cost[3].

On the hardware-side the "mouse" therefore relies at the moment on radio-transmission of the camera-pictures to a PC-host which does the processing. But we are on the way to use a Phytec TI320C50 DSP-board with a piggyback frame-grabber (the whole hardware is smaller than a cigarette-pack) to do the complete job on the robots.

## 4.1 The visual analysis

On the software side of the vision, the active vision paradigm is followed as few computation-power must be sufficient to do all processing. The active vision paradigm has several properties which makes it the ideal solution for the visual performance we are aiming at. The purposive and task-oriented design — meaning that only what is needed is computed —, the use of cues and attentional mechanisms and the lack of elaborate internal representations reduce the computational complexity. Reducing the need for computational power is most important when using low-cost, portable hardware. Other properties are the tolerance to errors, the dependence on recognition rather than reconstruction, the roots in biological vision systems and the link to behavior based systems.

Several parallel-working modules handle the visual perception, each module handles a certain cue or task. The input to each module is the unaltered

---

[3] A "mole" roughly consumes 5 Watt, whereas the power-consumption of todays super-computers lies in the range of 10 to 100 Kilo-Watt.

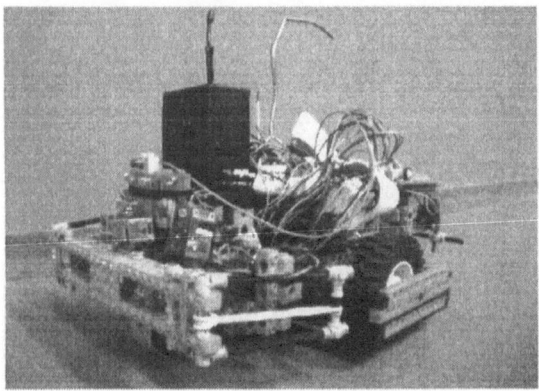

**Fig. 2.** The robot "mouse" with its camera at the front.

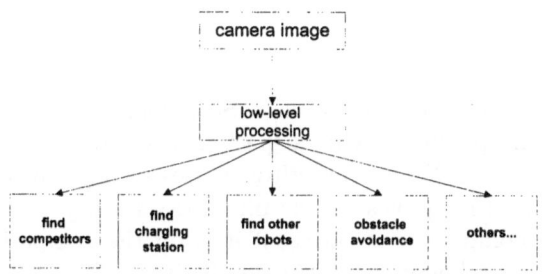

**Fig. 3.** The architecture of the vision system. Several modules each asynchronously handle a cue or task.

camera image or this image fed through some low-level analysis (typically some Sobel-edge detection and some basic smoothing). The architecture, with the in (simulated) parallel working modules, is shown in figure 3.

## 4.2 Seeing the competitors

The competitor-module has as tasks the detecting of competitors in the image and returning their position, a distance-estimate and their state (active or stunned). This problem can be tackled in a number of ways, e.g. by doing template matching or by using some neural network. We however use a very straightforward solution. All competitors are dark colored and simple thresholding of the image reveals all dark patches. All these patches are checked for their aspect ratio. If the aspect ratio is within a certain margin the same as the aspect ratio of a competitor, it is considered being one. This method works very fast and is very satisfactory for our purpose; the few times that it does not correctly recognize a competitor (this happens when two competitors appear as one dark

**Fig. 4.** A competitor as seen by the robot. A border is placed around the competitor, meaning that is has been recognized as active. To the right the charging station can be seen.

patch in the image) don't pose a problem. So recognizing the competitors is reduced to finding dark patches having a specific aspect ratio. Next, the state of each competitor is checked. Each competitor houses a lamp emitting modulated light —this is used by the "moles" to do photo-taxis using simple sensors— if the lamp is on, the competitor is active, if it is off the competitor is stunned (and the robot no longer homes in on the competitor). The competitor module checks for this light, and marks every competitor as active or not.

Estimating the distance to a competitor is simple. The competitors are cylindrical and look the same from every angle, so the distance to the observer is reversely proportional to their size in the image (see (1) where $a$ and $b$ are constants depending on the aspect ratio of a competitor and the units in which one would like to express the distance).

$$\text{distance} = \frac{1}{2}\left(\frac{a}{\text{width}} + \frac{b}{\text{height}}\right) \tag{1}$$

The module builds a radial depth map containing all parasites, their position in the image, their distance to the robot and their status (active or not). All this information is available to the behavior system of the robot, but of course not all this information is needed. The "align on competitor"-behavior only needs the position of the closest competitor alive; while the obstacle avoidance module needs the position of all competitors, no matter whether they are alive or not.

## 4.3 Seeing the charging station

The charging station has a bright white light, which is used by the "moles" to home in on the charging station using only their two light sensors. Since the white light is the most prominent feature of the charging station, it is also used

**Fig. 5.** The charging station, correctly recognized by the robot at a distance of 2m.

by the visual system to recognize the charging station. The module thresholds the incoming image for light colors and calculates the centroid. To avoid confusion caused by the overhead lights and their reflections on the shiny floor, the centroid is calculated in a restricted region in the middle of the image.

Recognizing the charging station with other algorithms (template matching, Hough-transforms,...) is extremely hard. This is mainly due to the charging station's irregular shape, to the fact that is can be viewed from an infinite combination of distances and angles and to the ever changing reflections of light in the ecosystem.

The floor of the ecosystem is flat and the camera viewpoint is always on the same level (the camera cannot tilt), so objects will appear at the same height in the image. This is called the *ground-plane constraint* and can be used to estimate the distance to objects, since farther objects will appear higher in the image. Calculating the distance to the charging station is however not easy. Using the position of the charging station in the image, or its size seems an obvious choice; but in practice the position and the size are very hard to determine because of the same problems that were experienced during recognition of the charging station. In a last attempt the station was fitted with black horizontal stripe, which is clearly visible and allows for a distance-measure (the height of the stripe in the image is inversely proportional to the distance).

## 4.4 Seeing obstacles

When the *ground-plane constraint* is respected, it is surprisingly easy to do obstacle avoidance; provided the floor of the environment is smooth (such as the uniform colored floor of the VUB AI-lab ecosystem). The lower part of the image always shows the floor, when doing edge detection this part of the image will show no or hardly any edges. Unless an obstacle is close enough to the robot to enter this lower image part. So walls, other robots and objects will be detected

**Fig. 6.** The "head", an agent-type forced to cooperate.

as soon as they come within a predefined distance to the robot. As a result the obstacle avoidance module returns the direction furthest away from all obstacles, the behavior system for the control of robot can use this to steer clear from all obstacles. This method has also been used by Horswill in his Polly robot [8].

### 4.5 Recognizing other agents

In some experiments (e.g. [24]) an agent needs to recognize other agents. Since robots are the only thing moving in the ecosystem, the "Find other robots"-module only needs to look for motion in the image not caused by the observer itself (this implies that agents have to move in order to be seen). One could use optical flow analysis for this, but since this is a computationally expensive way to handle the problem, difference images are used. This has one drawback over the optical flow-approach: the observer can not move while looking for other robots. This is however acceptable in our current experiments.

The "Find other robots"-module also monitors the ego-motion of an agent. This can be useful when the robot gets stuck. The robot bodies don't carry any wheel encoders, and the ego-motion information can be compared to the last motor actuation; if these don't match a retract-behavior should be stimulated.

## 5 Forced to cooperate: the "head"

The so-called "head" is the most recent agent-type of the VUB ecosystem. It is a very unconventional autonomous robot as it is not mobile. But though being forced to stay at one place, far away from the "food-source", it can preserve itself from starving. As we will see, it can provide useful information to other (mobile) animats. It does not do this for "free", but for a certain portion of the benefit in terms of energy. So, it is kind of "fed" by other robots in exchange for providing information.

The head consists of a camera mounted on a pan-tilt-unit (figure 6) and is equipped with quite some computing power. As it is bound to have strong vision-capabilities, the hardware needed is not feasible to be carried around by small or medium-sized robots. Being implemented on a sufficiently large mobile base, a "head" would be far from being competitive with the "moles" and "mice". Following the reasoning in section 3, the "head" is therefore conceptualized as immobile animat, which in addition adds substantially to the amount of heterogeneity in the system.

## 5.1 The capabilities of the "head"

Initially the head makes a 180 degree scan of the environment, building a radial depth map containing all competitors. After that it switches into a watch-dog mode, where it just randomly looks around the ecosystem. When its attentional mechanisms (which trigger on unusual motion) pick up something interesting, the focus is placed on that particular region. Since the only things moving in the ecosystem are the robots, the head will pick out a robot and start tracking it. The head will try to predict its path and will utter a warning (over the radio-link) if the robot comes close to a relevant object. Kuniyoshi et al. [9] built a system doing the same, but instead used a stereo head and more powerful hardware.

For the scanning of the ecosystem, the head rotates its pan-motor step-wise from left to right and analyses each image frame, meanwhile building the radial depth map of all things relevant. In the watch dog mode the head observes a part of the ecosystem for a few seconds, when nothing of interest is detected it performs a saccade (this is a quick repositioning of the camera, note the analogy with saccadic eye movements in mamals) and observes another part of the ecosystem. During the saccade the head is unable to see anything, due to its egomotion.

Finally, the tracking is done using difference images. As soon as the robot is about to move out of the image, the camera is repositioned to get to robot back into the middle of the image. We call this *saccadic tracking*. The head also continuously compares the position of the robot with the position of the competitors in the radial depth map, if a robot comes too close to a competitor it is warned using the synchronous radio link every robot carries.

The architecture of the head's visual analysis is basically the same as in figure 3, consisting of several modules running in parallel. All analysis and actions happen in real-time.

## 6  Conclusion

We presented the introduction of heterogeneity into to a robotic MAS. The system is motivated from an Alife point of view; the agents are linked together in their common need of resources, especially energy, to keep operational. The basic ecosystem consists of a charging station where mobile robots can re-fill batteries autonomously, competitors which establish a working task, and simple agents, the so-called "moles".

Given this homogeneous system, the implications of heterogeneity are discussed in an informal way. The notion of feature is introduced as capability of an agent to allocate resources within its ecosystem. An agent-type or "species" can be seen as collections of features. It is pointed out that features are grounded in the real world and therefore lead to an additional consumption of resources as well. It follows, that there is a trade-off between the benefit and the costs of a feature. Therefore, the design of new species has to take the environment and especially existing agent-types into account.

With these ideas in mind, the new species "mouse" and "head" are introduced. Especially, concrete technological choices and implementation details are presented and motivated.

The "mouse" is a kind of enhanced version of the "moles". In addition to simple sensors like the moles, the mouse is equipped with the feature vision. This allows it, among other capabilities, to recognize the charging station and the competitors. In doing so, two improvements over the moles are achieved. First, the mouse can see the charging station and the competitors over much longer ranges. Second, it can estimate their distances. Both improvements can be used to access energy more easily. The vision-processing and its hardware-needs are designed such, that their additional power-consumption are in relation to these benefits.

The "head" is the second new species in the ecosystem. It consists of a camera on a pan-tilt unit and is immobile. The immobility follows from its need of substantial computing-power. The head can help other animats by providing information: its is capable of tracking robots and it can issue helpful warnings. As it cannot get to the charging station itself, it relies on "selling" these informations to other robots.

**Acknowledgments**

Many thanks to Werner van Belle for writing parts of the "heads" software (environment scan and distance prediction to competitors). The robotic agents group of the VUB AI-lab is partially financed by the Belgian Federal government FKFO project on emergent functionality (NFWO contract nr. G.0014.95) and the IUAP project (nr. 20) CONSTRUCT. Robotic hardware development in the group is partially funded with the TMR-grant "Development of an universal architecture for mobile robots" (ERB4001GT965154). Parts of the robotics-research in the VUB AI-lab are used for grounding within the GOA2-project "Origins of Language and Meaning" (OZR/96/2156). Tony Belpaeme is a Research Assistant of the Fund for Scientific Research - Flanders (Belgium) (F.W.O.).

# References

1. Ross Ashby. *Design for a brain*. Chapman and Hall, London, 1952.

2. Hugues Bersini. Reinforcement learning for homeostatic endogenous variables. In *From Animals to Animats 3. Proc. of the Third International Conference on Simulation of Adaptive Behavior*. The MIT Press/Bradford Books, Cambridge, 1994.

3. Andreas Birk. Learning to survive. In *Fifth European Workshop on Learning Robots, Bari*, 1996.

4. Bruce Blumberg. Action-selection in hamsterdam: Lessons from ethology. In *From Animals to Animats 3. Proc. of the Third International Conference on Simulation of Adaptive Behavior*. The MIT Press/Bradford Books, Cambridge, 1994.

5. Rodney Brooks. Intelligence without reason. In *Proc. of IJCAI-91*. Morgan Kaufmann, San Mateo, 1991.

6. Robert Ghanea-Hercock and David P. Barnes. An evolved fuzzy reactive control system for cooperating autonomous robots. In *From Animals to Animats 4. Proc. of the Fourth International Conference on Simulation of Adaptive Behavior*. The MIT Press/Bradford Books, Cambridge, 1996.

7. Ian Horswill. Characterizing adaption by constraint. In *Toward a Practice of Autonomous Systems, Proceedings of the First European Conference on Artificial Life*. The MIT Press, Cambridge, 1992.

8. Ian Horswill. Polly: A vision-based artificial agent. In *Proceedings of the Eleventh National Conference on Artificial Intelligence*. AAAI, MIT Press, 1993.

9. Y. Kuniyoshi, M. Ishii, S. Rougeaux, N. Kita, S. Sakane, and M. Kakikura. Vision-based behaviors for multi-robots cooperation. In *IEEE Int. Conf. on Intelligent Robots and Systems*, 1994.

10. Long-Ji Lin. Self-improving reactive agents: Case studies of reinforcement learning frameworks. In *From Animals to Animats. Proc. of the First International Conference on Simulation of Adaptive Behavior*. The MIT Press/Bradford Books, Cambridge, 1990.

11. Pattie Maes. A bottom-up mechanism for behavior selection in an artificial creature. In *From Animals to Animats. Proc. of the First International Conference on Simulation of Adaptive Behavior*. The MIT Press/Bradford Books, Cambridge, 1990.

12. Maja J. Mataric. Designing emergent behaviors: From local interactions to collective intelligence. In *From Animals to Animats 2. Proc. of the Second International Conference on Simulation of Adaptive Behavior*. The MIT Press/Bradford Books, Cambridge, 1993.

13. D. McFarland and A. Houston. *Quantitative Ethology: the state-space approach*. Pitman Books, London, 1981.

14. David McFarland. What it means for robotic behavior to be adaptive. In Jean-Arcady Meyer and Stewart W. Wilson, editors, *From Animals to Animats. Proc. of the First International Conference on Simulation of Adaptive Behavior*. The MIT Press/Bradford Books, Cambridge, 1991.

15. David McFarland. Towards robot cooperation. In Dave Cliff, Philip Husbands, Jean-Arcady Meyer, and Stewart W. Wilson, editors, *From Animals to Animats 3. Proc. of the Third International Conference on Simulation of Adaptive Behavior*. The MIT Press/Bradford Books, Cambridge, 1994.

16. Jean-Arcady Meyer and Agnes Guillot. Simulation of adaptive behavior in animats: Review and prospect. In *From Animals to Animats. Proc. of the First International Conference on Simulation of Adaptive Behavior*. The MIT Press/Bradford Books, Cambridge, 1991.

17. Alexandros Moukas and Gillian Hayes. Synthetic robotic language acquisition by observation. In *From Animals to Animats 4. Proc. of the Fourth International Conference on Simulation of Adaptive Behavior*. The MIT Press/Bradford Books, Cambridge, 1996.

18. Lynne Parker. On the design of behavior-based multi-robot teams. *Advanced Robotics, 10 (6), pp.547-578*, 1996.

19. Luc Steels. The artificial life roots of artificial intelligence. *Artificial Life Journal, Vol 1,1*, 1994.

20. Luc Steels. A case study in the behavior-oriented design of autonomous agents. In Dave Cliff, Philip Husbands, Jean-Arcady Meyer, and Stewart W. Wilson, editors, *From Animals to Animats 3. Proc. of the Third International Conference on Simulation of Adaptive Behavior*. The MIT Press/Bradford Books, Cambridge, 1994.

21. Luc Steels. Discovering the competitors. *Journal of Adaptive Behavior 4(2)*, 1996.

22. Luc Steels. A selectionist mechanism for autonomous behavior acquisition. *Journal of Robotics and Autonomous Systems 16*, 1996.

23. Luc Steels and David McFarland. *Cooperative Robots: A case Study in Animal Robotics*. The MIT Press/Bradford Books, Cambridge, 1994.

24. Luc Steels and Paul Vogt. Grounding adaptive language games in robotic agents. In Phil Husbands and Inman Harvey, editors, *4th European Conference on Artificial Life*. The MIT Press/Bradford Books, Cambridge, 1997.

25. Toby Tyrrell and John E. W. Mayhew. Computer simulation of an animal environment. In *From Animals to Animats. Proc. of the First International Conference on Simulation of Adaptive Behavior*. The MIT Press/Bradford Books, Cambridge, 1990.

26. Barry Brian Werger and Maja J. Mataric. Robotic "food" chains: Externalization of state and program for minimal-agent foraging. In *From Animals to Animats 4. Proc. of the Fourth International Conference on Simulation of Adaptive Behavior*. The MIT Press/Bradford Books, Cambridge, 1996.

27. Gregory M. Werner. Using second order neural connections for motivation of behavioral choices. In *From Animals to Animats 3. Proc. of the Third International Conference on Simulation of Adaptive Behavior*. The MIT Press/Bradford Books, Cambridge, 1994.

28. Stewart W. Wilson. The animat path to ai. In *From Animals to Animats. Proc. of the First International Conference on Simulation of Adaptive Behavior*. The MIT Press/Bradford Books, Cambridge, 1991.

# Designing Organized Agents for Cooperation with Real Time Constraints

Michel Occello, Yves Demazeau, Christof Baeijs

LEIBNIZ/IMAG/CNRS
46, avenue Félix Viallet
38031 Grenoble Cedex FRANCE
{Michel.Occello,Yves.Demazeau@imag.fr,Christof.Baeijs}@imag.fr

**Abstract.** The aim of this paper is to present our approach for designing Multi-Agent Systems in the context of collective robotics, and more generally in the context of real time distributed artificial intelligence applications. The paper presents an agent model (ASTRO) especially adapted to a real time context and shows how the cooperation can be achieved with this model by integrating external organizations and interactions. A design methodology is introduced to build agents using social knowledge (interaction and organization). A platform is presented including software development tools supporting the approach.

## 1  Introduction

Collective robotics systems or real time distributed artificial intelligence applications can be simulated with multi-agent systems (MAS) or use the agent paradigm to model embedded decision making kernels. These applications need systems able to express the cooperation between entities, nevertheless they have to guarantee real time properties. The aim of the paper is to present our approach for designing MAS in this context. It presents an agent model (ASTRO) especially adapted to a real time context and shows how the cooperation can be achieved with this model by integrating external organization and interaction. In the first section, we study the impact of real time constraints on the agent notions. We then present the ASTRO model : an agent specialized for real time organized systems. We examine in the third section how we can realize cooperation using this agent model. Finally we give an insight of the implementation of the ASTRO architecture using the framework of the MASK platform for the development and the simulation of MAS.

## 2  Multi-Agent Systems in a Real Time Context

As a first approach, we could say that a global real time behavior for a Multi-Agent System (MAS) could be obtained by the sum of the real time behaviors of all the local agents. However, as the internal model of a single agent alone is not sufficient to describe the entire system, we need to take into account real time constraints in terms of interactions between agents, of perception and action in the environment, and of the organization of the whole MAS. As for the non real time BDI agent [11], we need to analyse the real time constraints, both in terms of :

  - internal functions of the agent (reasoning, decision, control) which guide the agent behavior,
  - and external functions (communication, perception, action) which manage the interactions and the organization. We study in the sequel the impact of real time constraints on models, we show that for real world applications, constraints have to be taken into account all along the design process of the agents, from the model to the implementation [15].

*Integration of Deliberative and Reactive Aspects.* In order to build real time multi-agent systems, we have to integrate in a single agent, cognitive capabilities (symbolic reasoning, social behavior) to ensure the best cooperation between agents, and reactive capabilities to follow the evolution of the environment as shown by Bussmann and Demazeau [3] :

- – An agent should be able to explicitly manipulate its knowledge about the universe,
- – Perception and communication has to be rich enough to maintain a coherent vision of the system's state,
- – Decision making and decision execution must be realized in time,
- – Identification of unpredicted events should be followed by an evaluation of internal changes,
- – Concurrency of the above described capabilities.

*Integration of Perception, Communication and Action.* The main aspect of the deliberative/reactive integration is to take into account the evolution of the *environment* and its impact on the reasoning process.
Approaches can be classified in two groups [12] based on
  - a hierarchical organization of perception, communication and reasoning capabilities where the system is a juxtaposition of deliberative and reactive capabilities rather than a real integration. In this type of models reactive layers ensure normal execution and trigger a deliberative layer to replan [8, 7].
  - a true integration of perception and communication in the decision making process with a functional decomposition of perception, communication and reasoning capabilities which cooperate continuously and asynchronously. The models possess evaluation mechanisms that modify the initial reasoning adaptatively [9, 1].
   For an optimal adaptation, it is necessary to address the problem of the focus to events because some events have no importance for the agent behaviour and others have to be treated with a high priority.

*Constrained Time Reasoning.* Actions but also reasoning processes, control of the internal state, and perception mechanisms have to be efficient according to deadline constraints. The modifications of the representation of the environment related to the perception of the environmental evolution introduce new goals for which the agents have to plan new actions that respond to the new situation. Reasoning capabilities have to respect deadlines imposed by the real time context; they must supply a plan in a limited time. Two approaches are proposed to solve this problem :

- using "any-time" algorithms refining an available solution continuously,
- using "design to time" algorithms applying different approximation levels guaranteeing a response in a fixed time.

Hybrid approaches may also be used.

*Social behavior.* The social behavior ensured by cognitive capabilities is in charge of the management of cooperation.

The *interaction* is the key element of the cooperation. In cognitive approaches (as in our case) the interaction media are messages. Time constraints introduce constraints on the message interpretation. The meaning of the response time to an event in AI systems has been studied in some other works [6, 10]. It is globally composed of a transmission delay, a reaction delay, an evaluation delay, an execution and processing delay. It clearly appears that, apart from the transmission delay (depending on perception capabilities) and the processing delay (depending on the actions), these delays don't concern behavioral characteristics but structural performances. Behavioral temporal characteristics of the messages' content address timed reasoning problems. The *organization* of the MAS generate dynamic problems from a temporal point of view if we consider auto-organization or re-organization mechanisms. In our case we make the assumption that the organizational framework is static i.e. that the organization is fixed when we define the MAS [2]. The organizational knowledge can constitute a usefull complement to the reasoning process if the organization is built according to the task dependencies including task efficiency information.

In summary, we will propose in the following sections to work towards a complete agent model, integrating :

- A symbolic representation of both the environment and the other agents,
- A sensing mechanism for both environmental and agent related events,
- A gradual adaptation of the behaviour in reaction to these events,
- A time constrained reasoning mechanism,
- A structure supplying parallelism and reactivity mechanisms in order to satisfy real time constraints.

# 3 ASTRO: A Model for a Real Time Agent

*A Model Integrating Deliberative and Reactive Capabilities.* The integration of deliberative and reactive capabilities is possible through the use of parallelism in the structure of the agents. Separating Reasoning/Adaptation and Perception/Communication tasks allows a continuous supervision of the evolution of the environment. The reasoning model of our agent is based on the Percep-

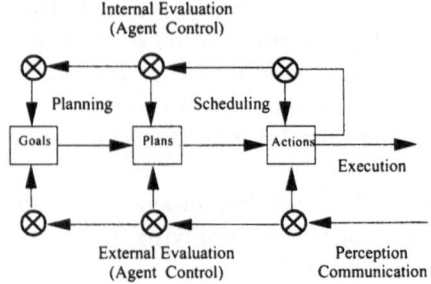

**Fig. 1.** Control Process of the Agent Model

tion/Decision/Reasoning/Action paradigm. The cognitive reasoning is thus preserved, and predicted events contribute to the normal progress of the reasoning process.

ASTRO can be presented as a "desintegrated agent" [13], where a functional decomposition in terms of capabilities provides a modular approach to the model. Decision modules evaluate the importance of the unpredicted events and have the obligation to place new actions or new goals in the internal state of the agent's reasoning. New goals imply the activation of the reasoning modules in order to partially or totally replan according to the importance of the event. New actions are directly placed on the agenda of actions in order to be executed in the specified delay. The agent control process can be explicited by fig. 1.

We now describe the different modules needed by such a deliberative/reactive agent :

**A Representation of the World.** The central part of the agent is its *world model*. This model comprises its knowledge about the environment, the internal states of other agents, and its own internal state. The proper internal state includes in particular the plans that are being executed or which the agent takes into consideration. This model is maintained by an interpretation process of the sensory data.

**Perception and Communication Modules.** Evolving in a real world, each agent has to integrate perception capabilities realized throught sensor devices. The knowledge about the environment is constructed by the perception modules. Other agents are "perceived" through communication modules. Agents can send information about their knowledge of the environment, their plans, their goals or their current state. Communication modules are probe loop events waiting for messages from agents. Emitters are considered as actions.

**Control Modules.** To ensure the reactivity of the agent, an evaluator continuously examines the world model. Agent control modules detect situations to which the agent needs to react, evaluate them, and decide to take the appropriate actions which may be to create, suspend, or kill goals, i.e. to change the context of the planning and executing process. The continuous supervision of the agent's situation ensures that the agent can react to unpredicted events at any time. The role of the perception evaluator is rather similar to the Perceptual Schema Controllers of the AuRa architecture of Arkin [1]. But additionally, we introduce same mechanisms to take into account the interactions with other agents using the interaction protocols proposed by [5]. Triggers and guards can be control loops observing the world representation or an evaluation function launched by the occurence of perception or communication events.

**Reasoning Modules.** The reasoning process consists of planning, scheduling and sequencing modules. Whenever a goal is created (or modified) a plan is searched that realizes the goal, this task is realized by planning modules.

We first describe the structure of the agent plans. Each plan has an identifier and an associated deadline. Execution of the plan requires the execution of several local goals. Each local goal has an identifier and an associated priority. The execution of each local goal may be accomplished by executing one of the several alternative actions. Each action has a duration (the execution time) and a satisfaction value associated to it.

Plans related to a given goal are stored in the part of the architecture concerning the internal state of the agent. The planner details the action in the order in which they will be executed. This process may be guided by hierarchical planning trying to infer the sequence of actions in a top-down fashion. In simple applications, we may assume that the agent has plans for every possibly encountered goals and possesses all the necessary actions; the planning process is in this case reduced to a fast pattern-matching algorithm. For complex applications, involving more social organizations, agents can negotiate with other agents about the actions they do not possess according to [17].

The purpose of the scheduling algorithm is to schedule the actions firstly, to meet the deadline and obtain a maximal utility value and a maximal satisfaction value.

To achieve this, we have established an algorithm informally described below.

1. For each plan, make a schedule of all the local goals taking minimal duration. If a deadline is violated at any stage, abort the plan and exit.
2. For a local goal with maximal priority from the remaining local goals :
   (a) if (action has already started) then go to step 2.
       else find the action with maximal satisfaction from the remaining actions
   (b) if(deadline is violated due to new action) go to step 2(a)
       else replace action with minimal duration by action with maximal satisfaction

Assuming that actions with higher satisfaction values require more execution time, this algorithm ensures that a schedule (if it exists) for meeting the deadline with a minimal satisfaction is made. The actions start executing according to the schedule. Then the highest priority local goal is taken and a higher satisfaction action is put in the schedule provided the action has not already started or if it does not violate the deadline. This is continued for the remaining local goals. Therefore, we provide to the agent an immediate schedule to start actions. Then try to improve it by further iterations, allocating more resources to actions having more value. At any time the algorithm has a current valid schedule. We thus have used a hybrid "any time"/"design to time" approach to the scheduling mechanism.

Actions are placed in an agenda. Furthermore, the scheduler has the possibility to include internal actions, such as replanning or the setting of *guard* and *trigger* modules, in order to be more adaptive at run time. Guards and triggers supply information about the current situation during the execution phase of the actions. Due to this technique, if the agent has to act fast, the scheduling and the execution of an incomplete plan can start before the planning process is completely finished. This ensures adaptation of the agent to the evolution speed of his environment and is necessary if the agent pursues several goals at the same time.

The committed actions are performed at the scheduled time by the action modules triggered by the sequencer module (executor).

# 4 Achieving Cooperation with ASTRO Agents

This section describes the approach we apply to build the MAS. The main point of interest is the way we integrate the interaction and organization knowledge in the reasoning process.

## 4.1 The docker robots experiment

The docker robots experiment concerns a simulation of three robots whose mission is to deliver objects into a box in a determined place in a warehouse. Conflicts occur during the execution of the tasks by the robots especially when two (or more) robots want to put down their objects at the same time. This is a case of shared resources management. We can solve the problem by exploiting organization and interaction knowledge, as we explain in the following sections.

## 4.2 An agent centered phase

**Scheduling of tasks.** The internal aspects of the agent's reasoning process, i.e. everything that the agent would have to do if it was alone, must be first analyzed. It is an agent centered phase. It consists of building plans. The set of actions needed to accomplish the mission must be defined. These actions are scheduled without taking external influences related to the work of other agents into account. This plan expresses the normal progress of the work the robot has to do for a given goal :

- A plan can include actions aiming to modify the environment or to acquire information from the environment,
- A plan can include actions for information exchange between agents. It is possible to define actions the agent will delegate during execution [17],
- A plan can include initialization of interaction processes with other agents.

In the case of our experiment we have a single plan constituted by a sequence of four tasks :

1. *Grasping*: The robot takes an object from a moving conveyor belt.
2. *Moving* : The robot moves to the right place in the warehouse.
3. *Putting down* : The robot puts the object down in the box.
4. *Returning to the grasping place* : the robot moves to its initial position.

**Representation of the environment and of the others.** In most cases, actions work with data available in the representation of the environment perceived by the agent. A definition of perception capabilities is then needed in order to maintain the state of the perceived environment. In the case of the delegation of tasks, agents that are able to execute these tasks must be specified. A representation of the competences of the other agents has to be built. In the experiment the state of the box (free or occupied) is the only information needed about the environment.

## 4.3 A society oriented phase

**Expression of the interation.** The interaction mechanisms using interaction languages with protocols is whitout any doubt the best adapted approach for our kind of agent model, even if other mechanisms can be used. ASTRO agents use the IL language [5] based on speech acts. IL uses protocols (represented as state diagrams as in our case on fig. 2).

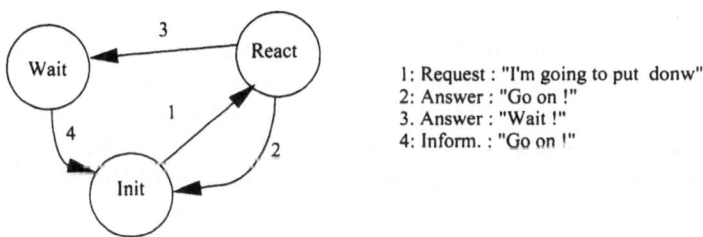

1: Request : "I'm going to put donw"
2: Answer : "Go on !"
3: Answer : "Wait !"
4: Inform. : "Go on !"

**Fig. 2.** The interaction protocol law (using the IL format) for the dockers robots

**Expression of the organization.** Our approach considers an external representation of the organization of the MAS. We use the RESO model [16] in which the organizational knowledge is expressed by a set of relations :

 - Relations of acquaintance : An agent who knows an other agent and has a representation of it, is in relation of acquaintance with this agent,
 - Relations of communication : Two agents who can exchange information are in relation of communication,

– Relations of subordination : An agent who is subordinated to another is in
relation of subordination with this agent.

These relations are quantified by a grammar established by Baeijs [2]. This
is a static expression of the organization in terms of social status.
In our application, we instantiate a hierarchical organization to the robots.
Robot3 is subordinated to Robot2 which is subordinated to Robot1 (Fig. 3).

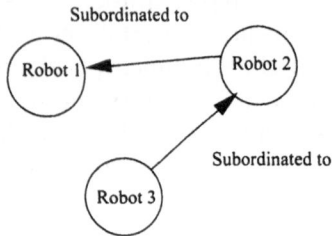

**Fig. 3.** The organization for the robots example

## 4.4 The integration of influences

The evolution of the environment, the interactions between agents and the organization of the MAS have an effect on the progress of the mono-agent plan
built at the beginning of the design process. The influences on the initial plan
have now to be analyzed. These influences are integrated in the agent decision
making process by building the capabilities of evaluation of the protocols' messages and of the perception detailled in the agent model as reactive elements.
Consequences of these influences are diverse :

– reaction of adaptation
– reaction on the plan progress (suspend action, suppress action ... )
– internal reaction : plan, new goal ...

In our experimental context, the progress of the protocol law will be constrained by the social status of the participant expressed in the organization.
Before passing a transition, the reasoning process has to consider the relation between the agents involved. For instance (Fig. 4), a subordinate (Robot2) doesn't
have to answer "Wait" to Robot1 if Robot1 ask him "Can I put down ?" (transition 1). Robot2 has to adapt his behavior even if he was going to put down
(the action must be suspended).

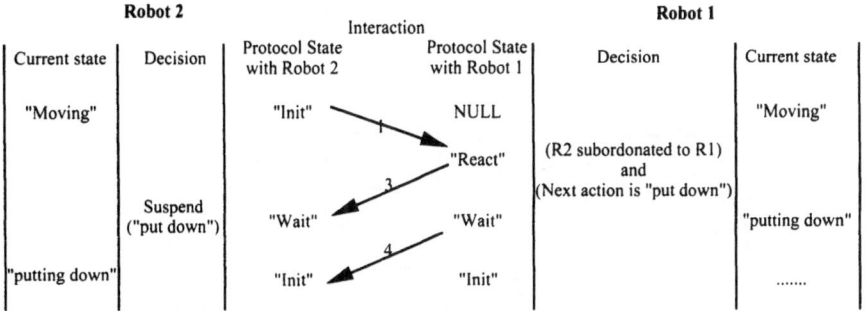

**Fig. 4.** Progress of the protocol

# 5 Building the Docker Robots Simulation Using the MASK Platform

## 5.1 The MASK Platform

The main goal of the MASK (Multi-Agent System Kernel) platform is to provide the multi-agent system designer a number of utilities packages embedded in a single software environment. The MASK platform (Fig. 5) is composed of packages covering different aspects of the multi-agent paradigm :

- The Agent package provides the user with pre-defined agent models or allows the definition of new ones.
- The Environment package provides the user functions to define and to work with a new (simulated) environment, or allows the user to use pre-defined-ones,
- The Interaction package is responsible for providing functions to use interaction protocols.
- The Organization package provides the user functions to establish the whole organisation of a multi-agent system.

Each of the four packages supplies the following features :

- Editors to define new models of the agent, environment, interaction and organization,
- Editors to build declarative parts of interaction and organization,
- Editors to build declarative parts of agents and to integrate interaction, environment and organization capabilities,
- Libraries to implement operational parts of the agent's capabilities

## 5.2 Building ASTRO Agents using the MASK ASTRO Toolbox

**The ASTRO Agent Architecture and Toolbox** The ASTRO agent model is implemented using a real time blackboard architecture. A parallel blackboard

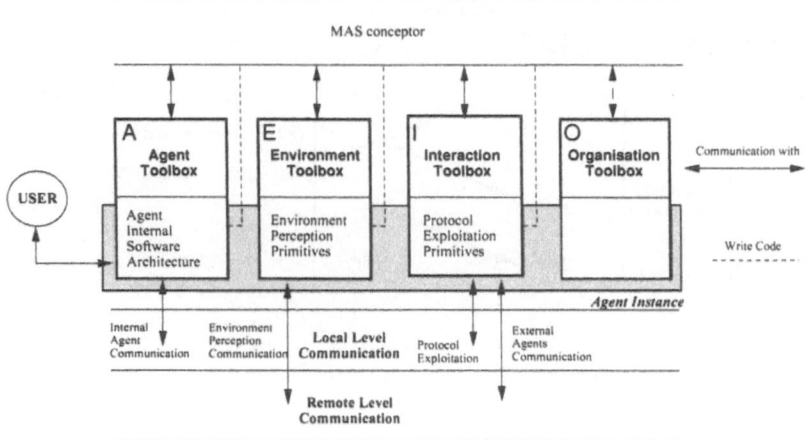

**Fig. 5.** The MASK Platform

architecture aims to express the inherent parallelism of the conceptual black-board model [4].

Modules react towards modifications of the blackboard, for their activations and inhibitions. They work on a local context which is a part of the blackboard data. *A domain blackboard contains domain data (used for problem solving). A control mechanism is in charge of the communication between modules and of the control of the management of the modules' activity.* Control data (summary of the state of the solution) are stored in a Control Blackboard managed by the control unit. Modules communicate with the control mechanism through event streams. The control unit sends a control stream to the modules. All the communications are managed by the controller. This blackboard control unit ensures stimulation and inhibition of the modules following their specifications.

The behavior of a module is described through its interactions with the control data, i.e. a representation of this behavior is given. This behavior is managed by message exchanges with the control unit. A module is integrated in the system by the specification of its behavior when faced to the blackboard data. An external specification of the behavior of the module can be expressed by an objective, preconditions of activation and interruption conditions. The control unit receives events from modules and emits control signals to them. Modules that have all their conditions validated are activated by an activation flow. Inhibition signals trigger exception processing in the modules. The control unit is application independent. We detailed and formalized this system in [14].

*Integration of the Model and the Architecture.* The different modules of the agent are organized according to the blackboard model described above. The model of the world constitutes the *domain blackboard*. All modules are managed by the control unit. This multi-modules approach allows a modular and independent description of each of the action and perception tasks in separate modules. The communication primitives can be of two types : internal communication (between the control unit and each module) and inter-agent communication (managed by the model of interaction).

A generic tool has been developed in C++ using UNIX communication libraries.

**Building capabilities using other Toolboxes** The ILAPI toolbox implements the IL language, it proposes a graphical interface to specify protocol laws, and primitives for the exploitation of the interaction knowledge. The RESO toolbox proposes a graphical interface to specify relations, and primitives for the exploitation of the organization knowledge. Finaly, the ASTRO Toolbox allows to build each capability of the agent by including primitives from ILAPI and RESO libraries. The external description of the protocol law and of the organization are built with the aid of the associated graphical tool and exploited by the agent modules using corresponding primitives.

The fig. 6 shows all the relations between elements.

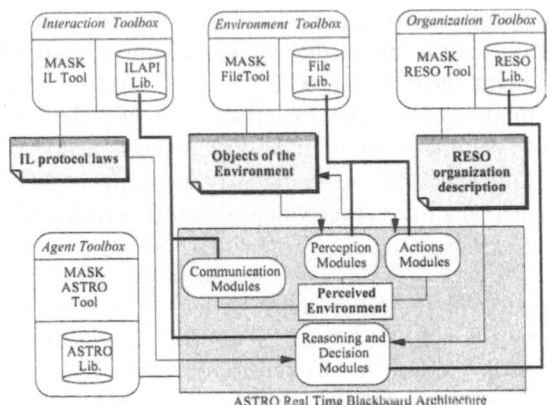

**Fig. 6.** Using MASK to build ASTRO Agents

## 6 Conclusion and Perspectives

*An Agent Model for Control* We have presented the agent concept as an integrator for heterogeneous complex devices and software. The evaluation of ASTRO shows that this agent model/architecture is viable for non-reflex decisions. The desintegrated structure with asynchronous modules allows to leave reflex decision in some modules whose internal control loops are directly related to the environment. Furthermore, the independence between modules allows the coupling of both simulated and physical control modes of activity.

*Viewing Parallel and Distributed Blackboards as Support Architectures for Real time Agents* The blackboard approach guarantees software flexibility and modularity in the implementation of Action and Perception/Communication tasks. The study of a parallel blackboard model specially adapted to reactivity brings a powerful support for problems involving both centralized representation of data (as cognitive agents) and reaction to unpredicted events (reactive aspects). A multi-agent system is in this case implemented as a distributed hybrid blackboard system, using a set of parallel blackboard systems.

The real time blackboard ASTRO agents are implemented on a UNIX workstation network using C++. This release uses a multi-processes local implementation of the internal architecture of the agent and a really distributed execution between agent entitites located on distinct hosts. An extension of the software is currently under development using the real time kernel PSOS + for hard real time problems.

*An Approach to Design MAS with ASTRO Agents* We detailled three phases to specify cooperation and integration of deliberative and reactive capabilities with ASTRO Agents. Our approach chooses to express MAS using the Agent, Interaction, Organization, Environment decomposition proposed by our team [5]. Organization and Interaction knowledge are global knowledge. Our contribution must be situated more at the design level than at the analysis level. The objective is to build operational agents. The experiment shows the application of the proposed approach to cognitive agents including reactive properties. It's typically an agent centered approach. We are currently working on a comparison with other approaches, such as a pure reactive approach (environment or organization centered), in order to propose an integrated methodology to design MAS.

*Integration in the MASK Platform* This unified design methodology will be entirely supported by the MASK development and simulation platform of multi-agent systems currently studied in our group. The ASTRO agent model is thus integrated as a toolbox in the MASK platform. ASTRO, ILAPI and RESO are operational toolboxes. Other Toolboxes are currently under study and being developped.

# References

1. R.C. Arkin and D. MacKenzie. AuRA: Principles and practice in review. *Journal of Experimental and Theoretical Artificial Intelligence*, 9(2), 1997.
2. C. Baeijs. Fonctionnalité émergente dans une société d'agents autonomes : Etude des aspects organisationnels dans les systèmes multi-agents réactifs. *PhD Thesis (In French), Institut National Polytechnique de Grenoble,* (To appear) 1998.
3. S. Bussmann and Y. Demazeau. An agent model combining reactive and cognitive capabilities. In *Proceedings of IEEE International Conference on Intelligent Robots and Systems - IROS'94,* Munchen, September 1994.
4. D.D. Corkill. Advanced Architectures: Concurrency and Parallelism. In V. Jagannathan, R. Dodhiawala, and L.S. Baum, editors, *Blackboard Architectures and Applications,* chapter II, pages 77–83. Academic Press, 1989.

5. Y. Demazeau. From cognitive interactions to collective behaviour in agent-based systems. In *Proceedings of 1st European Conference on Cognitive Science*, Saint Malo, France, april 1995.

6. R. Dodhiawala, N.S. Sridharan, P. Raulefs, and C. Pickering. Real-Time AI systems: a Definition and an Architecture. In *Proc. of the International Joint Conference on Artificial Intelligence - IJCAI 89*, 1989.

7. I.A. Fergusson. Toward an architecture for adaptative, rational, mobile agents. In E. Werner and Y. Demazeau, editors, *Decentralized A.I.* North Holland, 1992.

8. E. Gat. Integrating reaction and planning in a heterogeneous asynchronous architecture for mobile robot navigation. *SIGART Bulletin*, 2, 1991.

9. B. Hayes-Roth. An architecture for adaptive intelligent systems. *Artificial Intelligence*, 72(1-2):pp. 329–365, january 1995.

10. F.F. Ingrand and V. Coutance. Real time reasoning using procedural reasoning. Technical Report TR 93104, LAAS, Toulouse, France, 1993.

11. D. Kinny, A. Rao, and M. Georgeff. A Methodology and Modelling Technique for Systems of BDI Age nts. In W. Van de Velde and J. Perram, editors, *7th Workshop on Modelling Autonomous Agents in a Multi-Agen t World, MAAMAW'96*, volume LNAI 1038, pages 56–71. Springer-Verlag, 1996.

12. Jacek Malec. A unified Approach to Intelligent Agency. In M. Woolridge and Jennings N., editors, *Proceedings of ECAI-94 ATAL Workshop on Agent Theories, Architectures, and Languages*, volume LNAI 890, pages 232–244, Amsterdam, The Netherlands, August 1995. Springer-Verlag.

13. D. Moffat and N.H. Frijda. Where there's Will there's an agent. In M. Woolridge and N. Jennings, editors, *Proceedings of ECAI-94 ATAL Workshop on Agent Theories, Architectures, and Languages*, volume LNAI 890, pages 245–260, Amsterdam, The Netherlands, August 1995. Springer-Verlag.

14. M. Occello. Distributed and parallel blackboards : application to dynamic systems control in robotics and computer musics. *PhD Thesis Report (In French), University of Nice - Sophia Antipolis*, january 1993.

15. M. Occello and Y Demazeau. Modelling decision making systems using agents satisfying real time constraints. In *3rd IFAC Symposium on Intelligent Autonomous Vehicles*, Madrid, Spain, march 1998.

16. F. Scholastique. RESO : a tool to represent and exploit organizational knowledge. *Master of Science (In French), CNAM*, march 1998.

17. J.S. Sichman, R. Conte, and Y. Demazeau. A social reasoning mechanism based on dependence networks. In *Proceedings of ECAI'94 - European Conference on Artificial Intelligence*, Amsterdam, The Netherlands, August 1994.

# Tasking Robots through Multimodal Interfaces: The "Coach Metaphor"

**Luc JULIA**
STAR Laboratory
SRI International
333, Ravenswood Avenue
Menlo Park, CA 94025
Tel. +1 (650) 859 4269
Fax +1 (650) 859 5984
julia@speech.sri.com

**Abstract.** This paper presents multimodal interfaces to task and control multiple robots controlled by an agent-based architecture. For the past few years, SRI International have followed an approach based on the "Coach Metaphor". In sports or business, coaches are meant to apply predefined strategies to their teams, or, if something goes wrong, to find new means and plans during an ongoing game, so as to retask either the entire team or specific players. This is also the challenge facing a robot's operator. SRI's agent-based framework, the Open Agent Architecture™ (OAA), provides communication between the members of a team and the external world. The coach, or the robot's operator, who is an active member of the team, is provided with a multimodal interface that uses pen and voice. The analogy of a coach talking and drawing on a white clipboard representing the virtual world where the players are developing their game reinforces the metaphor. We present several interfaces specifically developed for SRI's robots, and we show an example (controlling robots on a soccer field) where the metaphor matches, one to one, the real world. To clarify our views, we will give an overview of the technologies in use, such as the agent architecture, the speech and gesture recognizers, and the robot controller.

## 1 Background

SRI International, has a long history of building autonomous robots, from the original Shakey through Flakey and more recently, the Pioneer class of small robots. Our Saphira architecture, an integrated sensing and control system for robotics applications, provides ways to represent tasks and plans, and tools to interpret sensor data, build and describe the environment [12]. In 1996, SRI won the AAAI "hold a meeting" event by putting together the latest results on real-time vision for robots, multimodal interfaces, and multirobots planning with an agent-based architecture [7]. The multirobots strategy opened new perspectives for several future projects: Because robots can now work as a team, operators need efficient ways to monitor them.

At the interface point of view, the AAAI contest innovation came from the ability of an operator to control and task robots through a multimodal, pen and voice, interface. Directly inspired from the Multimodal Map (MMAP) application [2], the Mapper (Figure 1) allows the user to interact with a dynamic map display through a natural combination of speaking, writing, and drawing directly on the map surface. New commands became available, such as

- *"What's the shortest path from here to the director's office"* + pen gesture
- *"Robot 1, follow this path"* + drawing with the pen

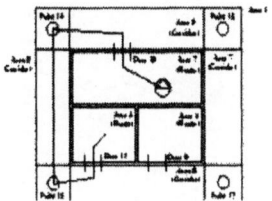

**Figure 1: The Mapper Interface**

In another ongoing project, aimed at sending autonomous robots with cameras onto the field (Small Unit Operations), we focus on the teleoperation of the robot and its onboard equipment. In addition to the MMAP/Mapper-like interface, the user is provided with a range of multimodal commands with which to operate any camera and device on any connected robot. Gesture driven commands on live video, vision algorithms, visual feedback, and direct manipulation are the key features of this interface (Figure 2).

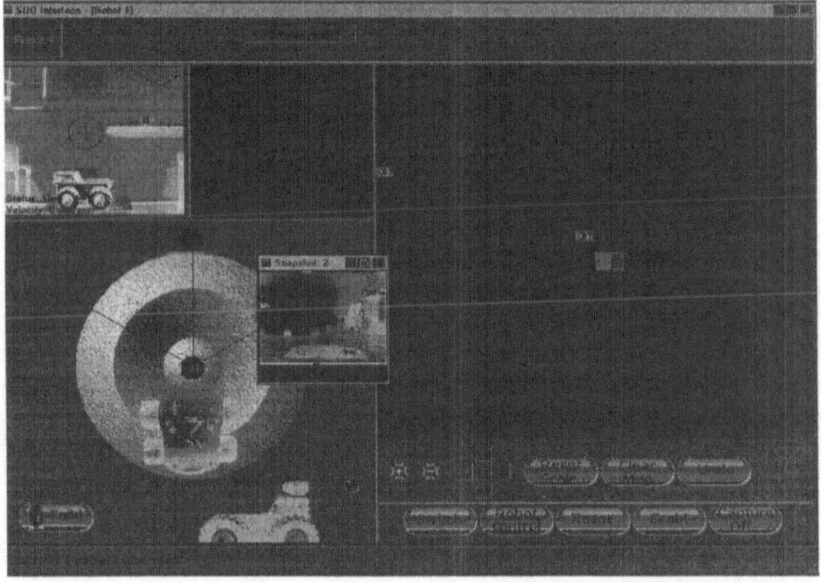

**Figure 2: The SUO Interface**

Nevertheless, robots still have autonomous capabilities, enhanced by object tracking and image processing algorithms developed in part for the MVIEWS[1] project [3]. In the current prototype, we have integrated several image functions, such as stabilization and extraction of selected regions, as well as two object tracking algorithms (Figure 3). Moreover, a geolocation component allows the mapping of objects from the 3D, video, world to 2D.

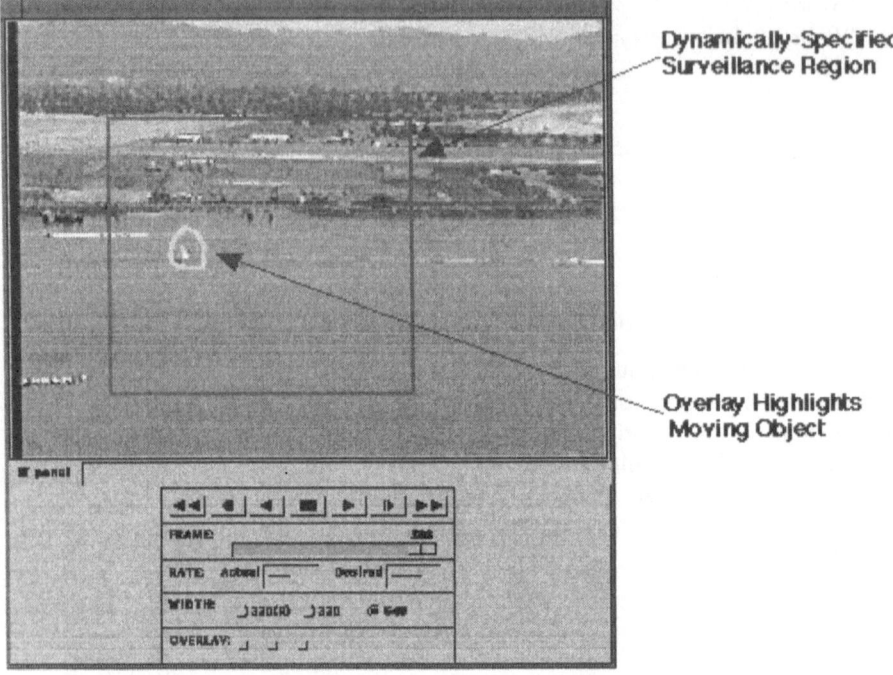

**Figure 3: Tracking in MVIEWS**

Commands like

- "If this object moves, notify me" + pen gesture on the video
- "If something appears in this area, alert me immediately" + pen gesture

allow the operator, in a robotics context, to send several team members onto the field without having to monitor each of them closely. The independence of each robot gives the operator more time to take care of the big picture, the strategy, and the opponent. The robot send an alert if a trigger is fired.

---

[1] http://www.erg.sri.com/projects/mviews

# 2 Open Agent Architecture

To manage the triggers, but more generally to glue together all the pieces, we use the Open Agent Architecture (OAA[2]), which was developed at SRI as a way of accessing many different types of information available in computers at different locations. Similar in objective to distributed object frameworks such as OMG's CORBA or Microsoft's DCOM, a distributed agent architecture such as the OAA can integrate components written in different programming languages[3] and running on different platforms[4]. However, OAA agents possess qualities beyond ordinary distributed objects. Agent interactions are more flexible and adaptable than tightly bound IDL[5] method calls in CORBA or DCOM, and can take advantage of parallelism and dynamic execution of complex goals. Instead of preprogrammed unitary method calls to known object services, an agent can express its requests in terms of a high-level logical description of what it wants done, along with optional constraints specifying how the task should be performed. This information is processed by one or more Facilitator agents, which plan, execute, and monitor the coordination of the subtasks required to accomplish the end goal.

The OAA has been used to implement more than 20 different applications, including

- Office automation and unified messaging
- Collaborative multimodal user interfaces
- Front ends and back ends for the Web
- Development tools for creating and assembling new agents with the OAA

Each OAA project can take advantage of the core services provided by the architecture as well as of the growing number of technologies now accessible through an agent interface. These services and technologies include speech recognition, natural language understanding, text extraction, multimodal fusion and reference resolution, reactive planning, virtual reality, image processing, Web-related technologies, user modeling, and collaboration tools. The core services of the OAA are implemented by an agent library, which has been ported to several different programming languages, working closely with a Facilitator agent, responsible for domain-independent coordination and routing of information and goals. These basic services can be classified into three areas: agent communication and cooperation, distributed data services, and trigger management.

---

[2] http://www.ai.sri.com/~oaa

[3] C, C++, Prolog, Lisp, Java, Borland Delphi, and Microsoft Visual Basic.

[4] UNIX (SunOs, Solaris, Lynx), Windows (3.1, 95, NT), all Java platforms.

[5] Specifies an object's methods by using a C++-like header file.

## 2.1 Interagent Communication Language

The Interagent Communication Language (ICL) provides the means for interaction among agents. When an agent wants to make a request of the agent community, it describes the goal it wants achieved as well as parameters specifying constraints on how the goal is to be accomplished. The request is sent to a Facilitator agent, which uses the declarative specifications it stores about each agent's capabilities, and the parameters defined for the incoming goal, to produce a fully specified execution plan detailing tasks for distributed agents to perform. The Facilitator agent is then responsible for monitoring and coordinating the execution of the plan, by routing requests (potentially to agents in parallel), collecting results, backtracking when subgoals fail, and finally providing the results to the requesting agent.

## 2.2 Data Management

OAA's distributed data facilities share much in common with the distributed goal resolution process described in the previous section. In the same way that agents register the tasks they are capable of performing, agents also declare descriptions of the data they manage. An agent can then add, delete, change, or query a data value, and this request will be automatically routed by the Facilitator agent to the appropriate agent or agents. Data declarations and functions also make use of the notion of parameter lists. In this case, parameters specify information about permissions, scoping, persistence, whether duplicate values are allowed, and so forth. Data parameters are also used provide synchronous collaboration features to OAA applications; the 'shareable' attribute determines whether a data value is synchronized among all participants of a distributed collaborative session.

## 2.3 Triggers

Triggers allow an agent, or set of agents, to monitor some potentially complex state in the world, performing an action if the trigger's test conditions become true. Triggers or rules exist in many commercial systems today; for instance, mail programs often allow the user to define actions (e.g., delete, archive, forward) to perform if email of a certain type arrives. However, in these systems, the action must be predefined and fixed. With OAA triggers, the action part of a trigger may be any compound task executable by the dynamic community of agents. As new, perhaps unanticipated, agents connect to the system at runtime, what the user can say and do literally changes. For instance, if a new fax agent suddenly becomes available, the user can now say or write «If email arrives…, fax it to Bill», even if this action had never been conceived of by the original developers of the application. Four triggers types are currently defined by the OAA:

- Data triggers: *«If the airline flight time changes…»*
- Time triggers: *«In ten minutes…», «every day from now until Christmas…»*
- Communication triggers: *«If any agent sends Msg…»*
- Task triggers: *«If mail arrives about…», «If this Web page changes …»*

Triggers are stored using the data management facilities, so they can be added, deleted, inspected, protected, and automatically distributed like any other database predicate.

Being able to make various programs work together just by plugging them into a set of other intelligent agents gives us a lot of flexibility. Now, our concern is how to add to this community another intelligent agent, the operator, with natural ways to communicate.

## 3 Multimodality, Speech and Gesture Recognition

Since the publication of Bolt's ground-breaking "Put-That There" paper [1], providing multiple modalities as a means of easing the interaction between humans and computers has been a desirable attribute of user interface design. Numerous user studies [14, 16] have shown that most subjects prefer combinations of spoken and gestural inputs. Not surprisingly, gestures provide a fast and accurate means of locating specific objects, while voice commands are more appropriate for selecting describable sets of objects or for referring to objects not currently visible on the screen. Many of these studies also attempt to enumerate and classify the relationships between the modalities arriving for a single command (complementary, redundancy, transfer, equivalence, specialization, contradiction) [13] in order to have a better understanding and a better model of the multimodal interaction and to design good human-computer interfaces.

In our approach, multiple modalities (handwriting, speech, pen gestures) may be entered simultaneously or in any sequential order, and merged to produce a command or request. To model interactions where blended and unsorted modalities may be combined in a synergistic fashion with little need for time stamping, we proposed a three-slot model [8] known as VO* V*, such that

- V (Verb) a word or a set of words expressing the action part of a command.
- O* (Object[s]) one or more objects to which the verb applies.
- V* (Variable[s]) one or more attributes necessary to complete the command.

Input modalities produced by the user fill slots in the model, and interpretation occurs as soon as the triplets produce a complete command. A multimedia prompting mechanism assists the user in fulfilling an incomplete command. This fusion makes use of the inherent parallelism of the OAA, with multiple agents competing and cooperating to resolve ambiguities arising during the interpretation process. The model does not take advantage of any particular modality, although we decided to add a priority mechanism to the spoken modality because of its semantic weight [11]. Such synergistic multimodal interfaces have been used in various OAA applications and experiments [15] that use speech, natural language, and pen recognition.

## 3.1 Speech Recognition

Speech recognition, along with natural language, is a huge component of the multimodal user interface. When it is possible to use any speech recognition product available on the market, we prefer the Nuance[6] recognizer, which is a real-time version of SRI's STAR laboratory continuous speech recognition system based on context-dependent genonic hidden Markov models (HMMs) [5]. SRI's DECIPHER™ technology recognizes natural speech without requiring the user to train the system in advance (i.e., speaker-independent recognition) and can be distinguished from the few other leading-edge speech recognition technologies by its detailed modeling of variations in pronunciation and its robustness to background noise and channel distortion. These features make the DECIPHER system more accurate in recognizing spontaneous speech of different dialects and less dependent on the idiosyncrasies of different microphones and acoustic environments.

## 3.2 Natural Language

In most OAA-based systems, prototypes are initially constructed with relatively simple natural language (NL) components, and as the vocabulary and grammar complexities grow, more powerful technologies can be incrementally added. It is easy to integrate different levels of NL understanding, depending upon the requirements of the system, just by plugging in an adequate engine. The available engines are two of our low-end NL systems: Nuance's template-slot tools and DCG-NL, a Prolog-based top-down parser. SRI's GEMINI [6] and FASTUS [8] are more powerful tools, used for complex NL tasks.

## 3.3 Pen Recognition and Annotation

Pen modality is used on different supports such as a map or live video to enter data such as handwriting and gestures.

The handwriting recognizer is provided by Communications Intelligence Corporation (CIC[7]), another SRI spin-off company, and plays an important role in labeling objects with out-of-vocabulary names, a task that is difficult for speech recognition systems. CIC is proposing fast letter-by-letter recognizers for Windows as well as for Unix and JOT, a Java-Graffiti like recognizer.

New gestures can easily be added to the common set (Figure 4) of the multistroke, multishape gesture recognizer developed in house. The meaning of each gesture can

---

[6] SRI spin-off : http://www.nuancecom.com

[7] http://www.cic.com/

be defined according to the domain of the application. Details on the algorithms can be found in [10].

**Figure 4: The Common Set of Gestures**

## 4 Conclusion: The "Coach Metaphor"

As an example that matches the real world, we describe an interface for tasking and controlling a team of robots playing a soccer game. The technologies and methods we have described all along this paper give us all the pieces needed for building such an application. Consider first what a good coach does in during a real soccer game:

*"The coach supervises, making sure that the team's basic strategy, defined in advance, is applied. If something goes wrong, the coach informs the players from the sideline, or at half-time, of a new plan. If a specific player does not respect the collective plan, the coach warns, teaches, scolds, and eventually gets the player out of the game".*

Then, consider an operator of our robots can do:

*"The operator chooses strategies according to the context, and watches the behavior of each team member. The operator then sends, on the fly, send a new behavior to a specific robot or to a subset of the team, or the operator sends a change in the whole strategy to all members of the team (no need for time-out or shouting from the sideline). If a robot fails, the operator can either replace it or give its position to another team member on the field according to the available resources. When the operator scolds a robot, a learning process may be involved."*

Moreover, as illustrated in Figure 5, the coach uses several modalities to convey the information to the players. He uses a clipboard to reference, to show objects in the shared world. Because he has a limited amount of time to make an idea clear, he uses the context (graphical, tactics) that every player understands.

**Figure 5: Coaching a team with a visual support**

With the multimodal interfaces we have developed, the soccer field is a map giving the operator the same flexibility as a real coach. We can even say that using pen and voice to task virtual players on the screen is easier than shouting orders, which then must be repeated from the sideline. This kind of interface provides easy ways to command and monitor the robots, by speaking predefined strategies, showing directions directly on the field by drawing simple gestures, or even combining both modalities to define new, complex behaviors. The visual feedback on the virtual screen, giving the position of each robot, allows the operator to have a better view and understanding of what is happening. The direct manipulation of virtual robots on the screen enables very fast communication with the real robots/players. Moreover, in the similar way we did in the SUO project, if each robot carries its own video camera, the coach can switch to a particular player's field of view and set goals by speaking and drawing on the video image (Figure 2).

The "coach metaphor" interface is superior to our earlier interfaces used to pilot and task robots. Although, the internal representation of the data is the same, the means to enter the data is much more efficient because the operator does not need to type complex commands and heavy descriptions in difficult formalisms. Moreover, the unique and unified representation and input for the whole team avoids complex data and device management.

Numerous previous studies showed the efficiency of multimodal, pen and voice, interfaces. In order to confirm those results, we are currently in the process of evaluating the MMAP application, and more generally multimodal interfaces, using a dynamic Wizard of Oz approach [4].

In future work, we hope to implement this specific application with an actual soccer team of robots. In the OAA configuration for the coach metaphor (Figure 6), the main pieces are the interface agents, the physical robots, and the shared databases. All the planning, learning, strategy, and world-representation used by Saphira are located the databases.

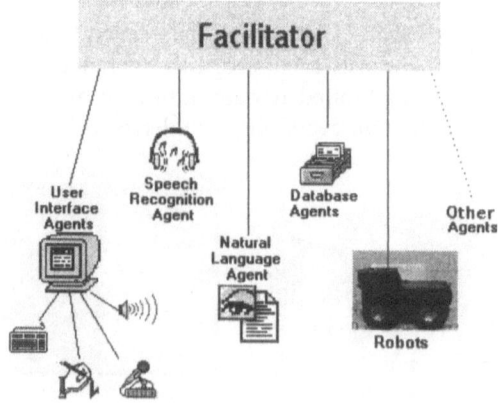

**Figure 6: OAA configuration for the "Coach Metaphor"**

# Acknowledgments

We would like to thank Adam Cheyer and Kurt Konolige, principal inventors of the agent (OAA) and robot (Saphira) architectures. Thanks also to Samir Gherbi, from the Ecole Polytechnique Fédérale de Lausanne in Switzerland, for the time he spent programming our interfaces.

# References

1• Bolt, R. Put-That-There: Voice and Gesture at the graphics interface. *Computer Graphics*, Vol. 14, Number 3, pp. 262-270, 1980.

2• Cheyer, A. and Julia, L. Multimodal maps: An agent-based approach. In *Proc. of CMC'95*, pp. 103-113, Eindhoven, The Netherlands, May 1995.

3• Cheyer, A. and Julia, L. MVIEWS: Multimodal Tools for the Video Analyst. In *Proc. of IUI'98*, pp 55-62, San Francisco, USA, January 1998.

4• Cheyer, A., Julia, L. and Martin, J.C. A Unified Framework for Constructing Multimodal Experiments and Applications. In *Proc. of CMC'98*, pp. 63-69, Tilburg, The Netherlands, January 1998

5• Digalakis, V., Monaco, P. and Murveit, H. Genones: Generalized Mixture Tying in Continuous Hidden Markov Model-Based Speech Recognizers. *IEEE Transactions of Speech and Audio Processing*, Vol.4, Num. 4, p 281, 1996.

6• Dowding, J., Gawron, J., M. Appelt, D., Bear, J., Cherny, L., Moore, R. and Moran, D. GEMINI: A natural language system for spoken-language understanding. *31st Annual Meeting of the Association for Computational Linguistics*. Pp. 54-61. Colombus, USA, 1996

7• Guzzoni, D., Cheyer, A., Julia, L. and Konolige, K. Many Robots Make Short Work. *AI Magazine*, Vol. 18, Number 1, pp. 55-64, Spring 1997.

8• Hobbs, J., Appelt, D., Bear, J., Israel, D., Kameyama, M., Stickel, M., and Tyson, M. FASTUS: a cascaded finite-state transducer for extracting information from natural-language text. in *Finite State Devices for Natural Language Processing* (E. Roche and Y. Schabes, eds.) MIT Press, Cambridge, USA, 1996.

9• Julia, L. and Faure, C. A multimodal interface for incremental graphic document design, In *Proc. HCI'93*, p 186, Orlando, USA, August 1993.

10• Julia, L. and Faure, C. Pattern recognition and beautification for a pen-based interface. In *Proc. of ICDAR'95*, pp. 58-63, Montreal, Canada, August 1995.

11• Julia, L. and Cheyer, A. Speech: A Privileged Modality. In *Proc. of EuroSpeech'97*, Vol. 4, pp. 1843-1846, Rhodes, Greece, September 1997

12• Konolige, K., Myers, K., Ruspini, E. and Saffiotti, A. The SAPHIRA Architecture: A Design for Autonomy. *Journal of Experimental and Theoretical AI*, Vol. 4, Number 0, pp. ?-?, ? 1997

13• Martin, J.C., Julia, L. and Cheyer, A. A Theoretical Framework for Multimodal User Studies. In *Proc. CMC'98*, pp. 104-110, Tilburg, the Netherlands, January 1998.

14• Mellor, B.A., Baber, C. and Tunley, C. In goal-oriented multimodal dialogue systems. In *Proc. ICSLP'96*, pp. 1668-1671, Philadelphia, USA, 1996.

15• Moran, D., Cheyer, A., Julia, L. and Park, S. Multimodal user interfaces in the Open Agent Architecture. In *Proc. of IUI'97*, pp. 61-68. Orlando, January 1997.

16• Siroux, J., Guyomard, M., Jolly, Y., Multon, F. and Remondeau, C. Speech and Tactile-Based Georal System. In *Proc. EUROSPEECH'95*, pp. 1943-1946, Madrid, Spain, 1995.

# Application of AOP for Modeling a Flexible Manufacturing Cell*

Carlos Rodríguez-Lucatero, Jesús Sánchez-V., Sergio Castillo-V.
ITESM, DIA **
Computer Science Department
Km. 3.5 Lago de Guadalupe, CP 52926, Atizapán, Edo.de Méx.
lucatero@campus.cem.itesm.mx
jsanchez@campus.cem.itesm.mx

Instituto Tecnológico y de Estudios Superiores de Monterrey, Campus Estado de México

Carlos Rodríguez-Lucatero, Jesús Sánchez-V., Sergio Castillo-V. ITESM, DIA Computer Science Department Km. 3.5 Lago de Guadalupe, CP 52926, Atizapán, Edo.de Méx. lucatero@campus.cem.itesm.mx jsanchez@campus.cem.itesm.mx

**Abstract:** We present in this paper a proposal for using the agent oriented programming paradigm as a tool for modeling agents that represent some of the elements of a real manufacturing cell. The main motivation in using this programming paradigm is that we want to give some kind of flexibility to manufacturing cells by exploiting the speech acts theory over which is based this paradigm.

**Keywords:** Multiagent Systems, Agent Oriented Languages, Scheduling, Flexible Manufacturing Cells, AGENT0, LALO, KQML.

## 1 Introduction

Roughly speaking, a manufacturing cell is a set of robots and numeric controlled machines that work together for the fabrication of some product. Manufacturing cells can be found in several industries such as car assembling, electronics and aeronautics. In the case of car industry, it is well known that they incur in very high production costs every time they need to change production parameters, because it implies reprogramming the robots, reconfiguration of the machines and tools adapted to the new kind of production, etc.

It is because of these reasons that it becomes necessary to conceive a manufacturing system that can respond to those changes without increasing manufacturing costs. To do this, we need to provide *flexibility* to manufacturing cells by implementing the following features:

---

* CONACYT (Consejo Nacional de Ciencia y Tecnología), y NSF Proyecto C016
** Instituto Tecnológico y de Estudios Superiores de Monterrey, Campus Estado de México

1. *Flexible Planning:* i.e. to have the capability to recognize plans under different formats and to adapt plans easily for having new goals.
2. *Reconfigurability:* i.e. to be able to change the manufacturing cell's layout without incurring in high costs.
3. *Detection and fault isolation:* i.e. to be able to detect, identify and isolate errors during the manufacturing cell operation.

We make a step towards implementing these features by designing a Parallel and Distributed Intelligent Controller (PARDICO) for manufacturing cells. The controller is *distributed* because the manufacturing cell elements are physically separated. It also must have some kind of *flexibility* so it is designed as a multiagent system. Finally, the controller is *parallel* to be able to support real-time operations and complex planning and execution schemes.

PARDICO's hardware and low-level software is described in detail elsewhere [SSA97] and shown in figure 1. The controller parallel's architecture is a tranputer network hosted by a PC and connected to each element of the cell by **RS-232** serial interfaces.

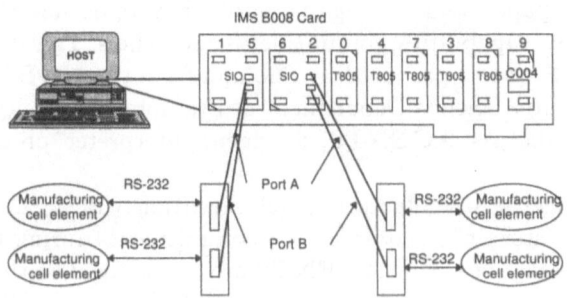

**Fig. 1.** PARDICO's architecture and interconnection network.

In this article we model a real manufacturing cell to get clues of how to design PARDICO's intelligent layer. We use the *Agent Oriented Programming* paradigm to model a manufacturing cell composed of two conveyors, a table, a milling tool, a welding tool, a Cincinnati robot, and two manipulator robots.

Each of these devices will be modeled by an agent that will be executed by a processor and will communicate by message passing with agents residing in other processors.

This paper is organized as follows. In section 2 we will describe briefly the AOP paradigm, and the language chosen for our research. In section 3 we discuss our approach for flexible planning in the manufacturing cell. In section 4 we describe the model that we generated for a typical manufacturing cell, and in section 5 we describe its implementation. Finally, we give our conclusions and plans for future work in section 6.

# 2  Agent Oriented Programming

One question that can be posed in this section is **why we have chosen the AOP** ? The reason of this choice is that this kind of language give us at the same time the modularity of the **OOP** approach and a *mental state* expressivity. Because of these features we think that the AOP approach is usefull for developing in a modular fashion the manufacturing cells model implementation.

From a developer's point of view the **AOP** can be seen as a specialization of the **OOP** (Object Oriented Programming), where the objects are more versatile, accept more specific types of messages as (*INFORM, REQUEST, DO*) and whose behavior is defined by mental states composed of:

- Beliefs about the other agents.
- Capabilities of the agent.
- Commitments with other agents.

In both, the **OOP** and the **AOP** paradigms, every object or agent receive messages. In the first case the objects respond always in the same way by the execution of a *method*, and in the second case the behavior of the object (agent) depends on its mental state.

In this part of the paper we are going to describe how to declare the structure of agents in terms of its capabilities, commitments and beliefs. This definition will be based on the **AOP** as was proposed by Shoham from the **NOBOTICS** laboratory of the Computer Science Department at Stanford University in [YS93]. Specifically, we will discuss **AGENT0**, an agent interpreter programmed in **COMMON LISP**.

Agents controlling the manufacturing cell execution will be informing, requesting, competing, assisting, and accepting commitments during their execution. The kind of messages used by **AGENT0** are constrained to those that are compositionally correct according to the *Speech Act Theory* [CP79]. This composition feature allows us to plan the communication actions to coordinate the activities of a set of agents for achieving a global goal without forgetting their own local goals. The speech acts in question must have the capability to modify the *mental states* of the interlocutors in such a way that we get a cooperative effect by communication. The main idea of **AGENT0** is to define agents in terms of mental states, using the elements defined in [YS93].

The logical framework consist in a quantified multimodal logic that enables us to make direct references over the time. This logic has three modalities: {Belief, Commitment, Capability}.

In **AGENT0** an agent is defined in terms of:

- A capability set (actions that can be performed).
- A set of initial beliefs and commitments.
- A set of commitment rules.

The key part that determines the agent's behavior is the set of commitment rules. Each commitment rule has a *message condition*, a *mental condition* and an *action*.

**AGENT0** was developed only as a prototype for showing the **AOP** principles, but it has not some important features as for instance a flexible planning system **AGENT0** has the following disadvantages:

- Agents can't make plans.
- Agents can't communicate in a higher level.
- The language does not really implement the associated logic.
- It is implemented in an interpreter, so it is unsuitable for real-time applications.

These weaknesses are related basically to the need to render the processing tractable. The truth maintenance of the modal logic knowledge bases is very expensive computationally, so **AGENT0** makes some assumptions to simplify it. For instance, the persistence of beliefs over time, simplicity of the formulas involved or the honesty of agents.

Another language very similar to **AGENT0** is **LALO** (Langage d'Agents Logiciel Objet, in french) by Gauvin, Marchal and Donne from **CRIM** (Centre de Recherche en Informatique de Montreal) [LALO97]. This language uses **KQML** (Knowledge Query Manipulation Language), has a compiler and generates **C++** code. The main reason that we have for using **LALO** is that it offers the same paradigm of **AGENT0** but additionally it gives the possibility of communicating under a standarized format (**KQML**) and it generates **C++** code that can be further compiled. This last point enable us to use **LALO** in a real time application.

**LALO** is based in an agent classes hierarchy from **reactive** to **deliberative**. The specification and the capabilities of an agent depend on its class. An agent has the following parts:

- Identification (Name of the class, name of the agent, etc.)
- Declaration of the agent tasks corresponding to its capabilities.
- Initial decisions (tasks, commitments).
- Initial Beliefs.
- The behavior, expressed as rules of three kinds: *to process the incoming messages, to execute tasks, to reason.*

In section 5 we will describe how we used **LALO** to model our system.

## 3   Flexible Planning and Reconfigurability

There are many approaches for defining what is a *plan*. Some definitions can be found in [RK91],[SIM63], [GRE69],[BUR77]. Planning in a manufacturing cell consists in:

- *Routing.-* selection of the sequence of machines that each part to be manufactured will visit.
- *Scheduling.-* the time and order in which each part will be processed in every element of the manufacturing cell.

Our planning approach is hybrid in the sense of being *centralized-distributed*. First, an off-line scheduler that works in a *centralized* fashion assigns tasks of the production schedule to each machine. Then, a *distributed* execution of the plan occurs in the multi-transputer controller.

During plan execution, there is a *Master* agent that plays the role of supervisor whereas the other agents are slaves. The Master agent has a global view of the system. Planning is made by the *Master agent* using case-based reasoning [SYC95]. To deal with contingencies, re-planning is done by agents using a *negotiation* scheme [KRA95].

In the current state of the development of our system there is a genetic algorithm based scheduler that assigns tasks to the Manufacturing cell elements.

One question that can arise here is **why we want to use a Speech Act based system ?**

The main motivation in using this approach is that we want to give some kind of flexibility to manufacturing cells by exploiting the speech acts theory over which is based this paradigm. The speech acts in question must have the capability to modify the *mental states* of the interlocutors in such a way that we get a cooperative effect by communication. These changes are recorded in the *mental states* in such a way that we can keep models of the other agents and communicate in a more directed fashion when the problems in the manufacturing cell appear. The advantage of this approach is to reduce the number of messages in the network by avoiding making a broadcast each time there is a failure or a conflict. A second advantage of this approach is that if we use languages based on **Speech Acts**, each agent can plannify its messages in such a way that change the *mental state* and *goals* of the other agents as was proposed in [CP79]. Generally speaking this feature can be exploited to improve the coordination between agents.

In the nest subsection we give a situation that illustrates when we can need to have flexibility.

## 3.1   Example of posible scenario

In our cell there are some elements whose functions can't be replaced because of their degree of specialization, and some others with overlapping capabilities. Therefore, in case of a failure, the functions of the latter can be performed by other elements. Those in the first case are the *Welding*, and the *Milling* elements. In the second case we have the *Robots* and the *conveyors*.

**Scenario:**

Suppose that $conveyor_1$ gets blocked while transporting the material $m_1$ through the **Cincinnati Arm**. The sequence of events we want our system to perform are:

1. $conveyor_1$ realizes that he is blocked, so he sends a message to the *Cincinnati Arm* to warn him that the material $m_1$ that he was waiting for at time $t_1$ is not going to be there.

2. The **Cincinnati Arm** believes that $Robot_1$ can reach the material $m_1$ on the $conveyor_1$ and move it to the desired location.

3. The **Cincinnati Arm** starts then to communicate with $Robot_1$ to get $m_1$ in the required position as soon as possible. The **Cincinnati Arm REQUESTS** him to get committed to move $m_1$ immediately.

4. Suppose that at that moment $Robot_1$ was doing something else, for instance feeding the $conveyor_1$ with more material. $Robot_1$ tries to find out if he can achieve the task. If the answer is **NO** then the **negotiation** is finished and the cell is stopped. If the answer is **YES** then $Robot_1$ sends a message to the **Cincinnati Arm** telling him that he is busy moving $m_2$ to the conveyor and that its higher priority is to do that. $Robot_1$ offers to do the task requested after it finishes in time $t_2$.

5. The **Cincinnati Arm** requests again to $Robot_1$ to leave $m_2$ on the table and move before $m_1$ because $conveyor_1$ is stopped.

6. $Robot_1$ accepts the request and gets committed to put $m_1$ at the required position at time $t_1 + k$.

Our main purpose in the present work is to show how we have agentified our elements using the **AOP** approach. As a further extension we want to implement a distributed scheduling system using a cooperative multiagent system.

One last question that we can ask here before talking about the implementation is **why agentify under AOP?**. The reason is that the *adscription of mental states* is *most useful* when applied to entities whose structure is very incompletely known or very complex as is the case for some elements of the manufacturing cell.

# 4 Manufacturing Cell Model

The manufacturing cell we are considering (figure 2) is composed of:

- A table used for storing primary material, in-process products and finished products.
- Two robots **IBM-7576** for heavy load manipulation.
- Two conveyors.
- A welding machine Maho-Graziano (CN) GR-300C (Lathe).
- A milling center Maho (CN) 700-S (Millmachine).
- A robot manipulator Cincinnati Milacron T3-374.

Each of these elements is controlled by an agent running in one of PARDICO's transputers. The heavy load manipulation robots are used to move parts from a table with materials to the conveyors back and forth. The function of the conveyors is to lend the material towards the processing facilities of welding and milling. The *manipulator robot* is the element of the cell that moves the materials located on the conveyors to and from the corresponding processing facilities towards the conveyors. The *processing facilities* are managed by a numerical

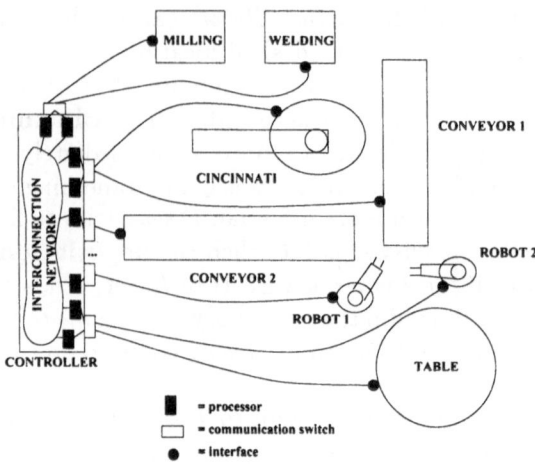

**Fig. 2.** A manufacturing cell and its controller.

control program. The execution of those programs has an average termination time that is part of the global cell model.

The master agent has initially a list of processes represented by a list of pairs $\{(r, p), \ldots\}$ where $r$ represents the resource (Welding or Milling) needed for the process, and $p$ is the identifier of the numerical control program that will be executed by $r$.

In every step of the plan, a material $m$ located in some element of the cell (table, $robot_i$, $conveyor_i$, manipulator, welding, or milling, where $i \in \{1, 2\}$) is processed to build product $x$. At the beginning, all of the materials are located on the table. As the production process advances, the materials are moved along the different elements. When finished, products are moved to the table.

These operations are performed cyclically until the job is finished. Our model takes into account a *technological data base*, containing the sequence of processes that must be performed to some material to transform it into a finished product.

We implemented every tool and robot controller as an agent, so our manufacturing cell software becomes a multiagent system composed of the following agents:

{ *Table, Robot$_1$, Robot$_2$, Conveyor$_1$, Conveyor$_2$, Arm, Milling, Welding, Supervisor* }

## 5  Implementation

### 5.1  Agents definition of the manufacturing cell

In this section we describe the manufacturing cell agents in terms of *Beliefs, Commitments, Capabilities and Commitment rules* as is done in the **AOP** paradigm. We will mention only a part of the definitions because of space limitations.

**Supervisor agents** We describe the mental state of the *Supervisor* agent in terms of **Capabilities**, i.e. the actions that can be performed by the agent, **Commitments**, i.e. the compromises that the agent can held with other agents and that must be coherent with its own capabilities, and **Beliefs** that are used to keep a model about the other agents and its own capabilities.

**Capabilities:** Execute-initial-plan, Monitoring-communications, Monitoring-processors, Monitoring-elements, Cell-state, Acceptation-test, Reconfigure-IMSC004, Execute-regressive-recuperation, Init-process, Suspend-process, Reboot-process, Finish-process, Replanning, Assign-tasks.

**Commitments:** Generate-initial-plan, Periodical-monitoring, Wait-contingency-agents-plan, Replanning, Start-process, Finish-process.

**Beliefs:** Agents-accept-tasks, Agents-start-execution, Agents-stop-execution, Agents-can-plannify, IMSC004-can-be-reconfigured, Processors-can-do-recuperation, Agents-answer-to-monitoring.

**Commitment Rules:**

```
IF (Generate-initial-plan= Ready)
THEN (Assign-tasks, Start-process)

IF (Cell-state = Finished-plan)
THEN (Finish-process)

IF (Periodical-monitoring = FAULT)
THEN (Fault-detected)

IF (Fault-detected AND Monitoring-processors = FAULT)
THEN (Suspend-process, Execute-regresive-recuperation, Reboot-process)

IF (Fault-detected AND Monitoring-communications = FAULT)
THEN (Suspend-process, Reconfigure-IMSC004, Reboot-process)

IF (Monitoring-elements = FAULT)
THEN (Suspend-process, Wait-contingency-agents-plan)

IF ((Wait-contingency-agents-plan = FINISHED)
    AND (Agent-can-plannify))
THEN (Reboot-process)

IF ((Wait-contingency-agents-plan = FINISHED)
    AND NOT (Agent-can-plannify))
THEN (Replanning, Assign-tasks, Reboot-process)
```

As can be seen from the definitions above this is a monitoring agent that has the right to assign tasks, stop or re-boot processes, replannify in case of a failure, etc. The behavior of the agents as well as the changements in its mental state are controled by the rules defined above.

**Table Agent Capabilities:** Verify-conveyor[i], Reserve-conveyor[i], Verify-utility[r], Reserve-utility[i], Retrieve-state-table, Receive-request-of-material[j], Material[j]-ready, Robot-transport-material-to-conveyor[i], Material[j]-on-table.

**Commitments:**

Accept-tasks, Accept-request-of-materia[j], Enable-departure-of-material[j], Enable-arrival-of-material[j], Record-material[j].

**Beliefs:**

Conveyor[i]-is-verifiable, Utility[i]-is-verifiable, Table-is-verifiable.

**Commitment Rules:**

```
IF (Receive-request-of-material[j]
    AND Material[j]-on-table
    AND (Verify-conveyor(1) = FREE
    AND  Verify-conveyor(2) = FREE)
    OR  ((Verify-conveyor(1) = FREE
    OR   Verify-conveyor(2) = FREE)
    AND  Verify-utility(r) = FREE))
THEN (Reserve-conveyor-free, Reserve-utility[r], Material[j]-ready)
```

As we can see from the definitions given above, this agent controls the operation of a stocking element at the manufacturing cell. The behavior of this agent is governed by the rules given above.

## 5.2 Implementation of our agents in LALO

In LALO, the agent program is divided into several sections:

- Identification (author identification, class name, agent name, etc.),
- Agent task declaration, that corresponds to the agent capabilities,
- Initial decisions,
- Initial beliefs (not used in our programs),
- Behaviors (rules).

The rules are divided by order of execution in: Message processing rules (RM), Task execution Rules (RE) and General Reasoning Rules (RG). In each loop for verification of rules conditions, just one rule is executed.

We use the following parameters in the messages exchanged by agents:

- u: Utility (Welding or Milling)
- p: Program number for execution in the utility.
- m: Sequential number assigned to each material in the table.
- r: Robot number (r=1 o 2)
- b: Conveyor number (b=1 o 2)

In this article we show only some portions of the agent table definition in LALO:

```
// +++++++++++++++++++++++++++++++++++++++++++++++++++++++++++++
// Agent Table definition in LALO (Mesa means Table in Spanish)
// File: Mesa.la
// Creation Date: 30-Sep-97
// Revision Date:  5-Oct-97, 11-Nov-97
// +++++++++++++++++++++++++++++++++++++++++++++++++++++++++++++

AUTHOR:          Sergio Luis Castillo Valerio;
EMAIL:           al130539@sunlab.cem.itesm.mx;
AGENT_CLASS:     MesaAgent;
AGENT_NAME:      Mesa;

TASKS:
        PRIVATE ATOMIC Verificar_mesa(m,for);
        PRIVATE ATOMIC Reservar_mesa(m,for);
        PRIVATE ATOMIC Reset_mesa();

DECISIONS:
        BEGIN_AT #begin: Reset_mesa();

BEHAVIORS:
// ---------------------- Message Rules ----------------------------
// (RMO)
   IF
      RECEIVED:
        achieve(sender: ?who, content:
                BEGIN_AT ?t: Verifica_mesa(m: ?m, for: ?who));
   THEN
      COMMITMENT_TO #myself:
        BEGIN_AT ?t: Verificar_mesa(m: ?m, for: ?who);

.....


// ---------------------- Execution Rules ----------------------------
// (REO)
   IF
      EXECUTING:
        Verificar_mesa(m: ?m, for: ?who);
   THEN SUCCEED;
....


// ---------------------- General Rules ----------------------------
// (RGO)
   IF
      BELIEF: AT ?now:
                Material_a_procesar_en_mesa(m: ?m, r: ?r, p: ?p, for: ?util);
   THEN
      COMMITMENT_TO #myself:
        BEGIN_AT ?now: achieve(sender: Mesa, receiver: Banda1,
```

```
            content: BEGIN_AT #now: Verifica_banda(b: 1, for: Mesa));
      BEGIN_AT ?now: achieve(sender: Mesa, receiver: Banda2,
            content: BEGIN_AT #now: Verifica_banda(b: 2, for: Mesa));

      BEGIN_AT ?now: achieve(sender:Mesa, receiver: ?util,
            content:
            BEGIN_AT #now: Verifica_utilidad(m: ?m, p: ?p, for: Mesa));

// (RG1)
   IF
      BELIEF: AT ?now: Bandas_libres(b: ?b, m: ?m, u: ?util, p: ?p, for: Mesa);
   THEN
      COMMITMENT_TO #myself:
         BEGIN_AT #now: achieve(sender: Mesa, receiver: Banda1,
                     content: BEGIN_AT ?now:
                           Reserva_banda(m: ?m, u: ?util, p: ?p, for: ))
....
```

## 5.3  Desk run for a process

In table 1 we show a desk run for a possible execution path for process $<$ $fresa, 2 >$ and material 1 ($m : 1$). The first column in table 1 is the sequence number corresponding to the time ordering of the events, the second one is the sender agent, the next one is the receiver agent, the fourth column indicates the triggered rule on sender, and the last one shows the contents of the message sent by the sender agent.

# 6  Results and Further Work

We have run a distributed simulation of the manufacturing cell operation using our definition of agents with **LALO** on a **PC under Windows95** and on **SPARC workstations** under **Solaris** operating system.

In the near future we will try to complicate the situations by simulating some failures to see how the manufacturing system degrades. Then, we will introduce several contingency schemes using **negotiation** to give the system the ability to continue operating even when there are some faulty elements. We will give now a possible scenario in our manufacturing cell, where a **negotiation** scheme can be used to solve a contingency.

## 6.1  Conclusions

The cooperative behavior shown in the last section can be achieved if we represent in our agents some beliefs about the cost of time in the negotiation process such that the *negotiation* converge quickly as is done in [KRA95]. For that we have to make some assumptions about the rationality of our agents and find out an optimizing function to each agent. In the case that the *negotiation* is not possible we ask a **Supervisor** to solve the conflict.

**Table 1.** Desk Run

| Seq | Sender | Receiver | Rule | Content |
|---|---|---|---|---|
| 1 | Mesa(M) | Banda1(B1), Banda2(B2), Utilidad Fresa (U:2) | RG0 | Achieve: Begin_at ?now: Verifica_banda(b:1..) Begin_at ?now: Verifica_banda(b:2..) Begin_at ?now: Verifica_util |
| 2 | Banda1 | Mesa | RG0 | Tell: Banda_libre(b:1,m:1,for:Mesa) |
| 3 | Fresa | Mesa | RG0 | Tell: Utilidad_libre(u:Fresa,m:1, ,p:2, for:Mesa) |
| 4 | Mesa | Banda1 | RG2 | achieve: Reserva_banda (m:1, u:Fresa,p:2,for:Mesa) |
| 5 | Banda1 | Mesa | RG3 | tell:Banda_reservada(b:1, m:1, for:Mesa) |
| 6 | Banda1 | Mesa | RG2 | tell:Banda_detenida(b:1, r:1 m:1, for:Mesa) |
| 7 | Mesa | Robot1 | RG6 | achieve:Translada_mat_a_banda (b:1, r:1, m:1, u:Fresa,p:2,for:Mesa) |
| 8 | Robot1 | —— | RM0 | commitment to #myself: Cargar_mat_a_banda(b:1,r:1, m:1,u:Fresa,p:2,for:Mesa) |
| 9 | Robot1 | Banda1 | RG2 | tell:Material_en_banda(m:1, u:Fresa,p:2,for:Mesa) |
| 10 | Banda1 | —— | RG4 | commitment to #myself: begin_at ?now: Desplazar_mat_a_utilidad (u:Fresa,m:1,p:2) |
| 11 | Banda1 | Brazo | RG7 | achieve:Carga_mat_a_utilidad (m:1,u:Fresa,p:2) |
| 12 | Brazo | Fresa | RM0 | achieve:Verifica_utilidad (u:Fresa,m:1,p:2) |
| 13 | Fresa | —— | RM0 | commitment to #myself: Verificar_utilidad (u:Fresa,m:1,p:2,for:Brazo) |
| 14 | Fresa | Brazo | RG0 | tell:Utilidad_libre(u:Fresa, p:2, for:Brazo) |
| 15 | Brazo | —— | RG0 | commitment to #myself: Cargar_mat_a_utilidad(m:1, u:Fresa,p:2,for: Fresa) |
| 16 | Brazo | Fresa | RG5 | tell: : Material_en_utilidad(m:1, u:Fresa,p:2,for: Fresa) |
| 17 | Fresa | —— | RG2 | commitment to #myself: Ejecutar_programa(p:2) |

In this paper we showed how to agentify the manufacturing cell elements. As a further work we will try to include in our agents *negotiation* features and experiment with different schemes to determine the ones that perform better in real situations.

Another future extension to our work will be the port of the **C++** code generated by **LALO** on a parallel transputer architecture using **parallel C**.

# References

[BUR77]  Burstall, R.M. and Darlington, J "A tranformation system for developing recursive programs". Journal of the ACM, 24(1):44-67, 1977.

[GRE69]  Green, C. "Application of theorem proving to problem solving". Proceedings of the First International Joint Conferece on Artificial Intelligence 1969, pag219-239

[SSA97]  J. Sánchez, M. Salmerón, J. Alarcón. "A parallel architecture to control flexible manufacturing cells". International Workshop on Parallel Computation and Scheduling in Computers (PCSC'97). Ensenada, México, 1997.

[SYC95]  Sycara Katia , Zeng Dajun and Kazuo Mayashita "Using Case-Based Reasoning to Acquire User Scheduling Preferences that Change over Time". IEEE 1995.

[YS93]  Yoav Shoham. "Agent Oriented Programming". Artificial Intelligence 60 (1993), pp. 51-92.

[KRA95]  Sarit Kraus, Jonathan Wilkenfeld and Gilad Zlotkin "Multiagent negotiation under time constrains". Artificial Intelligence 75 pag. 297-345, Elsevier Science, 1995.

[LALO97]  Daniel Gauvin, Hervé Marchal & Vince Delle Donne. CRIM. *http://www.crim.ca/sbc/english/lalo*, February, 1997.

[CP79]  Philip R. Cohen & C. Raymond Perrault. "Elements of Plan-Based Theory of Speech Acts". Cogn. Sci. 3 (1979) 177-212.

[RK91]  Elaine Rich & Kevin Night. "Artificial Intelligence". Mc Graw Hill, Second Edition, 1991.

[MS97]  Mirna Salmerón. "Diseño de una arquitectura de hardware reconfigurable utilizando transputers para el control de una CMF". Thesis for Master in Computer Science. ITESM-CEM, May 1997.

[LZ97]  Laura Zavala. "Comunicación tolerante a fallas en sistemas distribuidos". Thesis (in preparation) for Master in Computer Science. ITESM-CEM, 1997.

[SIM63]  Simon, H.A. "Experiments with a heuristic compiler". Journal of the ACM,10:493-506, 1963

[PAR96]  Van Parunak. "An introduction to Speech Acts and Dooley Graphs". http://www.iti.org/ van.

# Performance and Attention in Multi-Agent Tasks

Yiming Ye

IBM T.J. Watson Research Center
P.O. Box 704
Yorktown Heights, NY 10598, USA
yiming@watson.ibm.com *

**Abstract.** A well designed cooperation strategy for a task oriented multi-agent team is important as it can improve performance. A challenging research issue in cooperation concerns the extent to which an agent should pay attention to the actions and effects of other agents. In this paper, we address this issue in the context of an object search team. We first propose the concept of an activity window which captures an agent's view of the activities and effects of the team. Then we pinpoint some criteria that can be used to determine whether it is beneficial for an agent to put an action of the team into its window. Finally, we present experimental results to test these criteria.

## 1 Introduction

An agent is a computational system that inhabits dynamic, unpredictable environments. It has sensors to gather data about the environment and can interpret this data to reflect events in the environment. Furthermore, it can execute motor commands that produce effects in the environment. Multi-agent systems are computational systems in which several autonomous agents interact and work together to perform tasks or satisfy goals. Many researchers are building agents that can work in complex, dynamic multi-agent domains[8]. Such domains include virtual theater[1], realistic virtual training environments [5][6][8], RoboCup robotic and virtual soccer [4], among others.

Coordinating the actions of the agents is very important because an agent that considers the activities of other agents when forming its own plan is usually better able to choose actions that lead to outcomes that it favors. On the one hand, it is obviously not a good strategy for the agents of a cooperative multi-agent team simply to ignore each other, because the intended effects of one agent's action may already have been achieved by the actions of other agents. On the other hand, it is also not a good strategy for each agent to keep track of all

---

* The author would like to thank John Tsotsos, Demetri Terzopoulos, Allan Jepson, Chris Brown, Hector Levesque, and Eugene Fiume for their valuable comments on his Ph.D. thesis, and Eric Harley and the reviewers of CRW for their valuable comments on this paper.

the activities of the other agents, because the effort required might prevent the agent from doing useful work itself. For example, in the robotic soccer domain, ignoring teammates is ill-advised, but so is delaying action until one has complete knowledge of what all the other players are doing. A player that is going to shoot at the goal only needs to know its own surrounding and the situation around the opposing team's goalkeeper; while a goalkeeper only needs to observe the surrounding situation and the actions of the goal shooter in order to save the goal.

These observations motivate examination of how and to what extent an agent should consider the activities of other agents, and what factors are important in deciding local coordination strategy. Sen, Roychowdhury and Arora [7] study the effect of limited local knowledge on group behavior for the resource utilization problem where a number of agents are distributed between several identical resources. They conclude that an agent may benefit more from limited knowledge of the environment rather than complete global knowledge. Hogg, Huberman and Kephart [2] [3] analyze a similar problem and study the effects of local decisions on group behavior. They show that system parameters like decision rate can produce stable equilibria, damped oscillations, persistent oscillations, or chaos. Vidal and Durfee [9] present an algorithm for an agent to determine which of its nested, recursive models of other agents are important to consider when choosing an action.

In this paper, we address the issue of how and to what extent an object search agent should consider the activities of other agents during a multi-agent object search process — the process of searching for a 3D object in a 3D environment by a group of pan, tilt, and zoom cameras (or a group of robots). The goal of the team is to maximize the probability of detecting the target within a given time constraint. Given the real-world nature of the multi-agent object search task, the issues studied in this paper reflect many of the characteristics of other real-world tasks in a dynamic multi-agent environment.

## 2   The Multi-agent Object Search Team

In this section, we describe some concepts concerning the multi-agent object search system. These concepts are important for further discussions in the rest of the paper. We assume throughout the paper that there are in total $m$ search agents $a_1$, $a_2$, ..., $a_m$ available in the team.

The model of the search agent is based on Laser Eye - a pan, tilt, and zoom camera (Fig. 1(a)). The state $s_a$ of a search agent $a$ is uniquely determined by 7 parameters $(x_a, y_a, z_a, w_a, h_a, p_a, t_a)$, where $(x_a, y_a, z_a)$ is the position of the camera center, $w_a, h_a$ are the width and height of the solid viewing angle of the camera, $p_a, t_a$ are the the camera's viewing direction (Figure 1(c)). An operation $\mathbf{f}(a_i, s_{a_i}, r_{a_i}^{(j)})$ for agent $a_i$ entails two steps: (1) take a *perspective* projection image according to state $s_{a_i}$, and then (2) search the image for the target using the recognition algorithm $r_{a_i}^{(j)}$. We assume that each agent can have several recognition algorithms that can be used to detect the target; $r_{a_i}^{(j)}$ refers to agent

63

$a_i$'s $j$th recognition algorithm. The cost $\mathbf{t}(\mathbf{f})$ for an action $\mathbf{f} = \mathbf{f}(a_i, s_{a_i}, r_{a_i}^{(j)})$ gives the total time needed for agent $a_i$ to execute the action. It includes (1) time to manipulate the hardware to the state $s_{a_i}$ specified by $\mathbf{f}$; (2) time to take a picture using the camera on $a_i$; and (3) time to run the recognition algorithm $r_{a_i}^{(j)}$ specified by $\mathbf{f}$.

To encode the agent's knowledge about the possible target position, the search environment $\Omega$ is tessellated into a series of elements $c_i$: $\Omega = \bigcup_{i=1}^{n} c_i$ and $c_i \bigcap c_j = 0$ for $i \neq j$ (Fig. 1(b)). In addition, we introduce another "cell" $c_{\text{out}}$ to refer to the region that is outside the search region $\Omega$. Each cell $c$ is associated with a probability distribution $\mathbf{p}$ for each agent. The term $\mathbf{p}(a_i, c_j, \tau)$ gives the belief of agent $a_i$ regarding the probability that the center of the target is within cell $c_i$ at time $\tau$. The term $\mathbf{p}(c_j, \tau)$ gives the real target probability distribution at time $\tau$. Before the search process, $\mathbf{p}(c_j, \tau)$ and $\mathbf{p}(a_i, c_j, \tau)$ $(1 \leq i \leq m)$ are the same, but they may diverge during the search process.

To calculate the effects of applying a given action by a given agent $a_i$, we introduce the detection function $\mathbf{b}(a_i, c_j, \mathbf{f})$, which gives the conditional probability that agent $a_i$ will detect the target given that the center of the target is located within cell $c_j$, and the operation is $\mathbf{f}$. The value of $\mathbf{b}(a_i, c_j, \mathbf{f})$ can be obtained by transformation from a pre-recorded standard detection function for the recognition algorithm used by $\mathbf{f}$ [10]. Obviously,

$$P(\mathbf{f}) = \sum_{j=1}^{n} \mathbf{p}(a_i, c_j, \tau_{\mathbf{f}}) \mathbf{b}(a_i, c_j, \mathbf{f}) \ , \tag{1}$$

gives agent $a_i$'s belief on the probability of detecting the target if $\mathbf{f}$ is applied, where $\tau_{\mathbf{f}}$ is the time just before $\mathbf{f}$ is applied. The actual probability of detecting the target can be calculated by replacing $\mathbf{p}(a_i, c_j, \tau_{\mathbf{f}})$ of the above term with $\mathbf{p}(c_j, \tau_{\mathbf{f}})$.

For any agent $a_i$, its beliefs on the possible target positions $\mathbf{p}(a_i, c, \tau)$ (for all $c$) change over time as the multi-agent team perceives the world. Suppose another agent $a_j$ executes an action $\mathbf{f}$, then if agent $a_i$ does not care about the effects of $\mathbf{f}$, its belief will stay unchanged. Otherwise, its belief will be updated according to Bayes law:

$$\mathbf{p}(a_i, c_j, \tau_{\mathbf{f}+}) \leftarrow \frac{\mathbf{p}(a_i, c_j, \tau_{\mathbf{f}}) \left(1 - \mathbf{b}(a_i, c_j, \mathbf{f})\right)}{\sum_{k=1}^{n, \text{out}} \mathbf{p}(a_i, c_k, \tau_{\mathbf{f}}) \left(1 - \mathbf{b}(a_i, c_k, \mathbf{f})\right)} \ , \tag{2}$$

where $j = 1, \ldots, n, \text{out}$.

For a given agent $a_i$, one of the most important tasks during the search process is to select actions to search for the target. This action selection process is guided by the agent's knowledge $\mathbf{p}(a_i, c_k, \tau)$ — the agent always selects an action $\mathbf{f}$ with the maximum $P(\mathbf{f})$ (Formula (1)). Usually, there are a huge number of actions to be considered, most of which are not necessary. In [10], we develop a method that reduces the many possible actions to a limited set $\Pi[a_i]$ of actions that must be considered. Thus, during the search process, agent $a_i$ only needs to select the next action from $\Pi[a_i]$. The method is to calculate term (1) for each action in $\Pi[a_i]$. The first action that maximizes term (1) is selected.

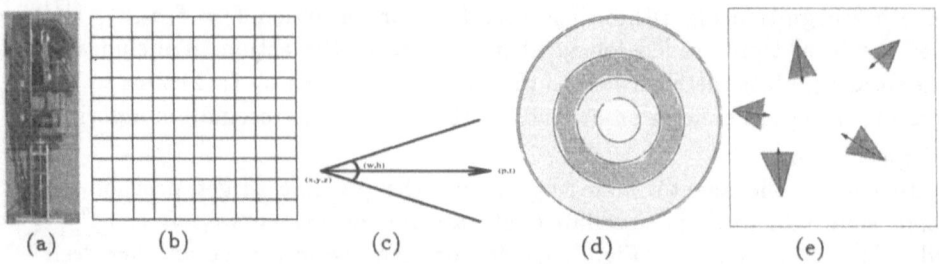

**Fig. 1.** *(a) The object search agent model: Laser Eye. (b) The tessellation of the environment. (c) The state parameters of an action. (d) The effective range (the dark layer) and the influence range (the dark and the light layers) (e) A scene of an object search team in an environment.*

The performance of the object search team is measured by the probability of detecting the target within a time constraint $T$ by the multi-agent object search team (Fig. (1)(e)). Suppose $\mathbf{F}_{a_i} = \left\{ \mathbf{f}_{a_i}^{(1)}, \mathbf{f}_{a_i}^{(2)}, \ldots, \mathbf{f}_{a_i}^{(N_{a_i})} \right\}$ is the set of actions actually selected by agent $a_i$ ($1 \leq i \leq m$) during the search process, where $N_{a_i}$ is the number of actions selected by agent $a_i$. Then $\mathbf{F} = \mathbf{F}_{a_1} \bigcup \mathbf{F}_{a_2} \bigcup \ldots \bigcup \mathbf{F}_{a_m}$ is the set of all the actions applied by the team during the search process. The total number of elements in $\mathbf{F}$ is $|\mathbf{F}| = N_{a_1} + \ldots + N_{a_m}$. Let $\Omega(\mathbf{f}_{a_i}^{(j)}) = \{c \mid b(a_i, c, f_{a_{i_1}}^{(j_1)}) \neq 0\}$ and $\Omega(\mathbf{f}_{a_{i_1}}^{j_1} \cdots \mathbf{f}_{a_{i_r}}^{j_r}) = \Omega(\mathbf{f}_{a_{i_1}}^{j_1}) \bigcap \cdots \bigcap \Omega(\mathbf{f}_{a_{i_r}}^{j_r})$. Let $\sum_{\mathbf{F}(\mathbf{f}_{a_{i_1}}^{(j_1)} \ldots \mathbf{f}_{a_{i_k}}^{(j_k)})}$ be the sum operation over all the different subset $\{\mathbf{f}_{a_{i_1}}^{(j_1)}, \ldots, \mathbf{f}_{a_{i_k}}^{(j_k)}\}$ of $\mathbf{F}$, where $1 \leq k \leq m$. Then the probability of detecting the target $P[\mathbf{F}]$ by the effort allocation $\mathbf{F}$ is calculated by the following formula [11]:

$$
P(\mathbf{F}) = (-1)^{1+1} \sum_{\mathbf{F}(\mathbf{f}_{a_i}^{(j)})} \left( \sum_{c \in \Omega(\mathbf{f}_{a_i}^{(j)})} \mathbf{p}(c, \tau_0) \mathbf{b}(a_i, c, \mathbf{f}_{a_i}^{(j)}) \right)
$$

$$
+ (-1)^{2+1} \sum_{\mathbf{F}(\mathbf{f}_{a_{i_1}}^{(j_1)} \mathbf{f}_{a_{i_2}}^{(j_2)})} \left( \sum_{c \in \Omega(\mathbf{f}_{a_{i_1}}^{(j_1)} \mathbf{f}_{a_{i_2}}^{(j_2)})} \mathbf{p}(c, \tau_0) \mathbf{b}(a_{i_1}, c, \mathbf{f}_{a_{i_1}}^{(j_1)}) \mathbf{b}(a_{i_2}, c, \mathbf{f}_{a_{i_2}}^{(j_2)}) \right)
$$

$$
+
$$

$$
\vdots
$$

$$
+ (-1)^{r+1} \sum_{\mathbf{F}(\mathbf{f}_{a_{i_1}}^{(j_1)} \ldots \mathbf{f}_{a_{i_r}}^{(j_r)})} \left( \sum_{c \in \Omega(\mathbf{f}_{a_{i_1}}^{(j_1)} \ldots \mathbf{f}_{a_{i_r}}^{(j_r)})} \mathbf{p}(c, \tau_0) \mathbf{b}(a_{i_1}, c, \mathbf{f}_{a_{i_1}}^{(j_1)}) \ldots \mathbf{b}(a_{i_r}, c, \mathbf{f}_{a_{i_r}}^{(j_r)}) \right)
$$

$$
+
$$

$$
\vdots
$$

$$+(-1)^{|\mathbf{F}|+1}\left(\sum_{c \in \Omega(\mathbf{f}_{a_1}^{(1)} \dots \mathbf{f}_{a_m}^{(N_{a_m})})} \mathbf{p}(c, \tau_0)\mathbf{b}(a_1, c, \mathbf{f}_{a_1}^{(1)}) \dots \mathbf{b}(a_m, c, \mathbf{f}_{a_m}^{(N_{a_m})})\right) . \qquad (3)$$

# 3 The Activity Window for a Given Agent

When an agent in a cooperative object search team (Fig. 1)(e) executes an action, it will also broadcast the parameters of the action (i.e., the viewing direction and the viewing angle size) to all the other members of the team (the communication time among agents for this purpose is small enough to be ignored). Thus, during the search process, an agent continually gets information on the action execution situations of other agents. The activity window $\mathbf{W}[a_i]$ for agent $a_i$ refers to the views of agent $a_i$ on the activities of the search team. By putting an action $\mathbf{f}$ executed by agent $a_j$ into the activity window of agent $a_i$ (represented as $\mathbf{f} \in \mathbf{W}[a_i]$) refers to the fact that agent $a_i$ updates its target distribution according to Formula (2) when $\mathbf{f}$ is executed by $a_j$. As we have discussed before, an agent $a$ selects actions based on its knowledge $\mathbf{p}(a, c, \tau)$ ($\forall c \in \Omega$). If it keeps track of every other agent's actions and updates its own knowledge (Formula (2)) accordingly, then its knowledge represents the true target distribution, thus it can select good quality actions during the search process. Otherwise, its knowledge will be different from the true distributions, thus it may not be able to always select good actions during the search process. Although it may seem that it is better for an agent to keep track of all the actions of other agents during the search process, this may not be a good strategy because it takes time to update the agent's knowledge. Thus, it is important for an agent in a multi-agent team to decide the extent to which it should heed the activities of other agents, or in other words, to decide the content of its activity window.

## 3.1 Factors that may influence team performance

As we have discussed above, good performance of the team depends on finding the proper balance between the benefit and the cost of updating an agent's knowledge based on what other agents are finding. Here, we pinpoint some factors that may influence this balance. Let $n_{update}$ be the number of actions in the activity window of agent $a_i$ which are heeded after the previous action is executed and before the beginning of the action selection process for the next action. Then there are three costs associated with an agent $a_i$'s action $\mathbf{f}$: (A) $t_{select}^{[i]}$, the time needed for the agent to select an action; (B) $n_{update} \times t_{update}^{[i]}$, the total time needed for the agent to update the environment for the $n_{update}$ actions in its activity window, assuming time $t_{update}^{[i]}$ for each; (C) $t_{execute}^{[i]}$, the time needed for agent $a_i$ to execute an action. If the cost $t_{execute}^{[i]}$ of executing an action is considerably higher than the update time $t_{update}^{[i]}$, then it is worth putting more actions in the activity window. Then more accurate knowledge about the target distribution is used and higher quality actions are selected. However, if $t^{[i]}_{execute}$ is considerably lower than $t^{[i]}_{update}$, then it is not worth spending time to keep track of the activities of other agents — better to devote the time to executing more actions.

Let $T$ be the time used for search; $n_i$ be the total number of actions applied by agent $a_i$ within time $T$; and $n_{ij}$ be the number of actions of agent $a_j$ that are heeded by agent $a_i$. Then for a team in which every agent keeps track of all the actions of other agents, the following relations hold:

(A) $n_{ij} \leq n_j$ $(j \neq i)$ and $n_{ii} = n_i - 1$.

(B) $n_i(t^{[i]}_{select} + t^{[i]}_{execute}) + (n_{i1} + \ldots + n_{im})t^{[i]}_{update} \leq T$ (for $1 \leq i \leq m$).

## 3.2   Determining the contents of the activity window

The important task of deciding which actions should be put in to the activity window is is difficult, because both the benefit, (which is not obvious sometimes), and the cost must be considered. Here, we discuss some criteria that can be used to determine whether it is beneficial to attend to an action of another agent.

Clearly, there may be certain agents in the multi-agent environment whose activities have no effect on set $\Pi[a]$ of potential actions for agent $a$. These irrelevant activities should not be considered by agent $a$, because there are no benefits. In the following, we study which agents' activities are irrelevant to agent $a$. We first explain some concepts. For a given camera angle size $\langle w, h \rangle$, there is an *effective range* corresponding to a spherical layer surrounding the camera (Fig. 1(d)) such that if the target is within this layer, the possibility that it be detected by a correctly directed action with size $\langle w, h \rangle$ is high. The actions in $\Pi[a]$ with size $\langle w, h \rangle$ are partly determined by this layer (refer to [10] for details). There is also an *influence range* corresponding to a larger spherical layer surrounding the camera (Fig. 1(d)), such that if the target is outside this range, it cannot be detected by an action with camera angle size $\langle w, h \rangle$ [10]. The outer radius of the influence range is the *influence radius* and is represented as $\mathbf{R}_a(\langle w, h \rangle)$. The intersection of the viewing volume of a given action $\mathbf{f}$ with the influence range for the angle size $\langle w, h \rangle$ of $\mathbf{f}$ defines the *influence volume* $\Omega(\mathbf{f}) = \{c \mid \mathbf{b}(a, c, \mathbf{f}) \neq 0,$ where $\mathbf{f} \in \Pi[a]\}$ (Section 2). An agent $a$ can have different zoom factors and thus different influence ranges. The smallest angle size that can be achieved by agent $a$ produces the largest influence radius, denoted $\mathbf{R}_a$. If the target is outside $\mathbf{R}_a$, then no matter how $a$ adjusts its state parameters, it will not be able to detect the target.

The following theorem and Properties can be used by agent $a$ to select its activity window during the team search process.

**Theorem:** *Suppose* $\mathbf{f}$ *is an action applied by agent* $a_j$ *during the search process. If* $\Omega(\mathbf{f}) \bigcap \Omega[a_i] = \emptyset$, *then there is no benefit in putting* $\mathbf{f}$ *in the activity window* $\mathbf{W}[a_i]$ *of agent* $a_i$. *In other words, the actions selected by agent* $a_i$ *will not be influenced whether* $\mathbf{f}$ *belongs to* $\mathbf{W}[a_i]$ *or not.*

**Proof:** Suppose $\mathbf{f}^*$ is the next action selected by agent $a_i$ when $\mathbf{f}$ is not included in $\mathbf{W}[a_i]$. Then we have: $P(\mathbf{f}^*) = max\{P(\mathbf{f}') \mid \mathbf{f}' \in \Pi[a]\}$ and $\mathbf{f}^*$ is the first such action chosen during $a_i$'s action selection process.

Now suppose that $\mathbf{f}$ is put in $\mathbf{W}[a_i]$. We need to prove that after the probability updating process of $a_i$, $\mathbf{f}^*$ will also be selected as the next action to execute. Let $\mathbf{p}(a_i, c, \tau)$ (for all $c \in \Omega$) be the target distribution when $a_i$ selects $\mathbf{f}^*$ in the situation that $\mathbf{f}$ is not in $\mathbf{W}[a_i]$; $\mathbf{p}'(a_i, c, \tau)$ be the target distribution after the probability updating process when $\mathbf{f}$ is in $\mathbf{W}[a_i]$; and $P'(\mathbf{f}')$ be the calculated probability of detecting the target by $\mathbf{f}'$ when $\mathbf{f}$ is in $\mathbf{W}[a_i]$. Then we have

$$P'[\mathbf{f}^*] = \sum_{j=1}^{n} \mathbf{p}'(a_i, c_j, \tau) \mathbf{b}(a_i, c_j, \mathbf{f}^*)$$

$$= \sum_{j=1}^{n} \frac{\mathbf{p}(a_i, c_j, \tau)(1 - \mathbf{b}(a_i, c_j, \mathbf{f}))}{\sum_{k=1}^{n, out} \mathbf{p}(a_i, c_k, \tau)(1 - \mathbf{b}(a_i, c_k, \mathbf{f}))} \mathbf{b}(a_i, c_j, \mathbf{f}^*)$$

$$= \sum_{c \in \Omega(\mathbf{f})} \frac{\mathbf{p}(a_i, c, \tau)(1 - \mathbf{b}(a_i, c, \mathbf{f}))}{1 - P(\mathbf{f})} \mathbf{b}(a_i, c_j, \mathbf{f}^*)$$

$$= \sum_{c \in \Omega(\mathbf{f})} \frac{\mathbf{p}(a_i, c, \tau)}{1 - P(\mathbf{f})} \mathbf{b}(a_i, c_j, \mathbf{f}^*) = \frac{P(\mathbf{f}^*)}{1 - P(\mathbf{f})} \geq \frac{P(\mathbf{f}')}{1 - P(\mathbf{f})} = P'(\mathbf{f}'). \qquad (4)$$

Thus, $P'(\mathbf{f}^*) = max\{P'(\mathbf{f}') \mid \mathbf{f}' \in \Pi[a]\}$. Since the algorithm selects $\mathbf{f}^*$ as the next action to execute when $\mathbf{f}$ is not included in $\mathbf{W}[a_i]$, it will also select the same $\mathbf{f}^*$ as the next action when $\mathbf{f}$ is included in $\mathbf{W}[a_i]$. $\square$

The above theorem is important because it states criteria for determining whether it is beneficial to include an action in an agent's activity window. However, it is not very convenient to use. The following properties give ways of conveniently using the theorem. Let $d_{ij}$ be the distance between agent $a_i$ and agent $a_j$.

**Property A** If $d_{ij} \geq \mathbf{R}_{a_i} + \mathbf{R}_{a_j}$, then it is not necessary for agent $a_i$ to consider actions executed by agent $a_j$.

**Property B** Let $\mathbf{f}$ be an action executed by agent $a_j$ with visual angle size $\langle w, h \rangle$. If $d_{ij} \geq \mathbf{R}_{a_i} + \mathbf{R}_{a_j} \langle w, h \rangle$, then it is not necessary for agent $a_i$ to put $\mathbf{f}$ into its activity window.

## 4  Experiments

A $2D$ simulation of a multi-agent object search system is implemented to test the influence of the activity window on the performance of the multi-agent object search system and to examine various factors that should be considered in deciding the content of the activity window. The system is implemented in C on IBM RISC System/6000. The search environment is a $2D$ square with size $175 \times 175$ as shown in Figure 2(a). There is an obstacle in the environment. We assume that each camera agent has only one fixed camera angle size $40°$. We also assume that the detection functions for each agent are the same. Figure 2(b) shows the detection function. It is obvious that the radius for any agent is 50. The time needed to select an action by an agent is determined by the size of the search region, the number of the candidate actions to be considered, and the speed of the machine used to run the code. The time needed to update the environment if an agent wants to incorporate an action's effects into its knowledge (that is, to put an action into its activity window) is determined by the size of the environment and the speed of the machine used to run the code. In our experimental setting, the average time needed to select an action is 21" seconds, the average time needed to update the environment is 7" seconds.

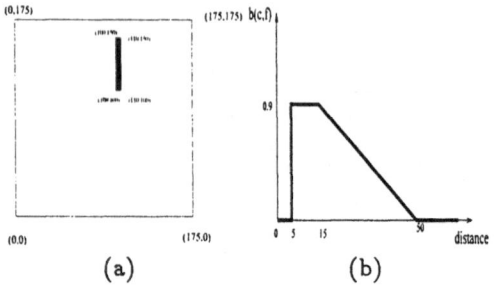

(a)                                            (b)

**Fig. 2.** *(a) The 2D environment with size* 175 × 175. *There is an obstacle bounded by* 100 ≤ $x$ ≤ 110 *and* 100 ≤ $y$ ≤ 150 *within the environment. (b) The value of the detection function.*

The first set of experiments tests the influence of the action execution time on the performance of the team when ($i$) each agent considers all the activities of other agents and ($ii$) when each agent ignores any activities of other agents. Figure 3(a) gives the scenario of the experiments. The initial outside probability $p(c_{out})$ is 0.05; the initial probability $p(c)$ for any element $c$ within the shaded area (bounded by 10 ≤ $x$ ≤ 165 and 45 ≤ $y$ ≤ 55) is 0.000179; the initial probability $p(c)$ for any element $c$ other than the shaded area is 0.000026. There are two agents within the environment. Their positions are (90, 15) and (91, 15) respectively. From Figure 3(b), we can see that for action execution times of 1" and 1.5", the strategy of paying attention to the other agent's activities does not perform as well as the strategy of ignoring the activities of the other agent. This situation changes gradually as the actions become more and more expensive, or in other words, as the action execution time becomes longer and longer compared to the action selection and the environment updating time. This illustrates that an agent should consider paying attention to the activities of other agents only when its action execution cost is high.

The second set of experiments test the criteria proposed in Section 3.2. The search environment and the initial target distribution are the same as before. But the positions of the two agents are changed to (35, 15) and (135, 15), respectively. The action execution time is 15" for the agent at (35, 15), and 15.5" for the agent at (135, 15). Since the distance between the two agents is equal to the sum of the influence radius of the two agents, the conditions in Property A hold. Thus, there is no benefit for one agent to keep track of what the other is learning. In

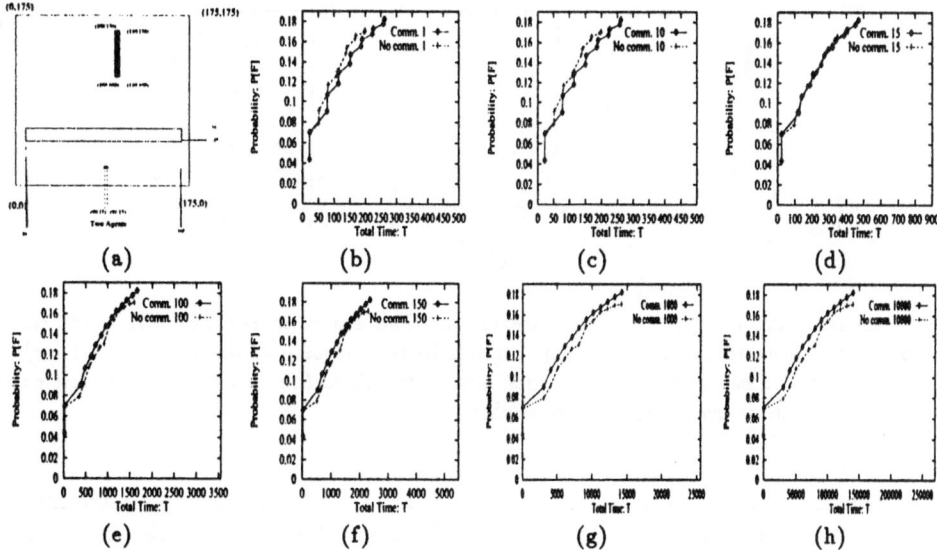

**Fig. 3.** *(a) The experimental scenario. (b)(c)(d)(e)(f)(g)(h) The probability of detecting the target P[F] versus the time constraint T for different execution times of the two agents, where execution times are: (b) 1", 1.5"; (c) 10", 10.5"; (d) 15", 15.5"; (e) 100", 100.5"; (f) 150", 150.5"; (g) 1000", 1000.5"; (h) 10000", 10000.5". In the figure, "Comm." refers to the strategy where each agent attends to the other agent's activities and "No Comm." refers to the strategy where each agent ignores the activities of other agents.*

this set of experiments, the actions and their sequence is exactly the same for the two strategies (communication and no communication). However, the start execution time for the actions with the same index in the two sequences are different, because in the communication strategy more time must be spent to update the environment. Figure 4(b) gives the delay in the execution starting time as a function of the action index of the team for the strategy of attending to the communications. Figure 4(c) compares the performance of the two strategies. The result is the same as one would predict from the theory in Section 3.2: in this case ignoring the communication between agents is better than attending to it.

The third set of experiments test the benefits of selectively controlling the content of the activity window based on the results in Section 3.2. There are 5 agents in the environment (Figure 5(a)). Two of them are on the left with positions $(45, 15)$ and $(46, 15)$ and action execution times 7.1" and 7.4", respectively. Three agents are on the right with positions $(155, 15)$, $(156, 15)$, and $(157, 15)$, and action execution times 7.5", 7.6", and 7.7", respectively. Based on Section 3.2, it is beneficial for agents on the left to attend to each other and for agents on the right to attend to each other, but it is not useful for any agent on the left (right) to listen to any agent on the right (left). The experimental results

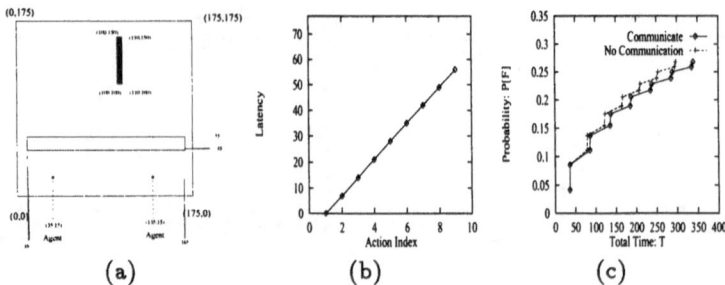

**Fig. 4.** *Testing the activity window criteria. (a) The environment. (b) The delay in action execution time when each agent attends to the activity of the other agent. (c) Comparison of the performance when each agent attends to the other agent's activities and the performance when they ignore each other.*

are shown in Figure 5(b). We can see that the strategy of selectively controlling the content of the activity window based on Section 3.2 is much better than the strategy in which each agent keeps track of all the activities of the team.

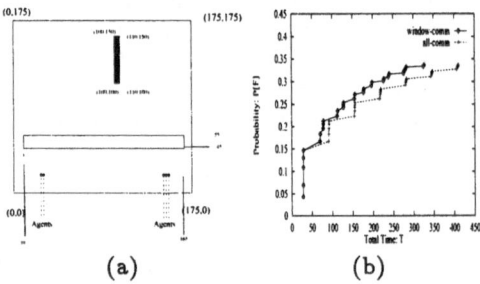

**Fig. 5.** *(a) The search environment. (b) Comparison of performance when the content of the activity window is selectively controlled based on Section 3.2 (window-comm) and the performance when each agent attends to all the activities of the team (all-comm).*

## 5   Conclusion

This paper addresses the issue of attending to knowledge acquired by the activities of fellow agents in a multi-agent domain — how much and to what extent should an agent care about what teammates are doing. The concept of activity window is proposed to represent the view of an agent on the activities and effects of its teammates. We study several factors that may influence performance when selecting the content of an agent's activity window in the context of an object search team, and we propose several criteria in this regard. Experimental results are presented which support our criteria and show the influence of the content of an agent's activity window on the team performance. We believe that some of the analysis presented in this paper can be applied to other multi-agent domains.

# References

1. B. Hayes-Roth, L. Brownston, and R. Gen. Multiagent collabration in directed improvisation. In *Proceedings of the International Conference on Multi-Agent Systems (ICMAS-95)*, 1995.
2. T. Hogg and B. Huberman. Controlling chaos in distributed systems. *IEEE Transactions on Systems, Man, and Cybernetics*, 21(6), 1991.
3. J. Kephart, T. Hogg, and B. Huberman. *Dynamics of computational ecosystems: implications for DAI*. Distributed Artificial Intelligence, Volume 2, Research Notes in Artificial Intelligence, Pitman, 1989.
4. H. Kitano, M. Asada, Y. Kuniyoshi, I. Noda, and E. Osawa. The robot world cup initiative. In *Proceedings of IJCAI-95 Workshop on Entertainment and AI/Alife*, 1995.
5. K. Pimentel and K. Teixeira. *Virtual reality: through the new looking glass*. Windcrest/McGraw-Hill, Blue Ridge Summit, 1994.
6. A. Rao, A. Lucas, D. Morley, S. M., and M. G. Agent-oriented architecture for air-combat simulation. Technical Report Technical Note 42, The Australian Artificial Intelligence Institute, 1993.
7. S. Sen, S. Roychowdhury, and N. Arora. Effects of local information on group behavior. In *Proceedings of Second International Conference on Multi-Agent Systems*, pages 315–321, Kyoto, Japan, 1996.
8. M. Tambe and P. Rosenbloom. Resc: An approach for real-time, dynamic agent tracking. In *Proceedings of the International Joint Conference on Artificial Intelligence*, 1995.
9. J. Vidal and E. Durfee. Recursive agent modeling using limited rationality. In *Proceedings of the International Conference on Multi-Agent Systems (ICMAS-95)*, 1995.
10. Y. Ye. *Sensor Planning for Object Search*. PhD thesis, Department of Computer Science, University of Toronto, Toronto, Canada, January 1997.
11. Y. Ye and J. K. Tsotsos. On the collaborative object search team: a formulation. In *Distributed Artificial Intelligence Meets Machine Learning, Lecture Notes in Artificial Intelligence Vol. 1221, Gerhard Weiß Ed.*, pages 94–116, 1997.

# *Cirta*: An Emergentist Methodology to Design and Evaluate Collective Behaviours in Robots' Colonies

Ouiddad Labbani-Igbida[1], Jean-Pierre Müller[2], and Alain Bourjault[1]

[1] Laboratoire d'Automatique de Besançon, UMR CNRS 6596,
olabbani@ensmm.fr, bourjault@ensmm.fr
[2] Institut d'Informatique et Intelligence Artificielle, Neuchâtel,
muller@info.unine.ch

**Abstract** An emergentist approach is necessary as long as we want to synthesise collective systems producing at the macro-level more than the sum of the micro-level individual capabilities or exhibiting new behaviours. In this paper, we introduce the methodology *Cirta* to design and evaluate emergent collective behaviours in robots' colonies and which consists (1) in specifying the conditions under which a desired macro-level behaviour is realised by micro-level interactions, (2) in instanciating the generic conditions in concrete interactions and (3) in evaluating the probabilities of occurrences. The different phases of *Cirta* are illustrated using a collective foraging behaviour. This paper is about the third stage where we use a stochastic model by Markov's chains to characterise the dynamics of each minirobot of the ecological niche and the dynamics of tracks and we prove that the global behaviour corresponds to chains' formation of minirobots connecting the nest to a food source.

## 1 Introduction

Facing the complexity and the misunderstanding of emergent phenomena, it seems natural to call for experimental methods to design collective behaviours. The non linear dynamics of interactions of the minirobots with the real world prevents reductionist methodologies from dealing with such phenomena. This situation becomes even more an issue when we want a multi-agent system to exhibit qualitatively new behaviours at the collective level or quantitatively perform more effectively.

Usual approaches in designing cooperative robots are bottom-up ([10], [14]). Based on intuition and experiments, they proceed by endowing the robots with a set of simple behaviours and according to what happens in the real world, they fit the behaviours to obtain a desired task. Among the top-down approaches, Collinot [2] proposes a methodological framework named *Cassiopee* using *role* assignment. A very popular approach in multi-agent systems is to use a knowledge level theory based on mental states such as beliefs, desires and intentions, to specify what knowledge based agents should be. But the way these theories could be derived into tractable and implementable systems is not clear despite some attempts like [15]. Finally, we can find attempts to globally describe the collective behaviours using statistical mechanics or dynamic systems like in [13].

It is generally admitted that there is no obvious way to derive individual properties from global behaviours, nor to ensure coherent and efficient collective behaviours from individual ones. In most cases, the derivation of the macro world properties remains a groping procedure and the evaluation of the macro world properties trials and errors procedure. Given individual properties, some authors like Kelly ([5]), Parunak ([12]) and Holland ([4]) propose a set of principles that a collective multi-agent system must match to exhibit interesting emergent behaviours. Ferber ([3]) proposes the effectiveness in working in groups and the existence of conflicts' resolution mechanisms as minimal criteria in the evaluation of groups' performance. Balch ([1]) also adopts a diversity metric to evaluate collective robots' performance inspired by the concept of Shannon's information entropy.

The main problem resides in defining the right level of description of the collective behaviours in order to understand or derive them from individual behaviours and conversely. How to design and specify the individual skills and their interactions (the micro level), giving a desired collective behaviour? And how to evaluate the performance of the global macro system, giving the micro world of its constituents? To answer these questions, we propose the methodology *Cirta*, which is based on a positive definition of the emergence concept and is organised in three stages. The first stage deals with the first part of the problematic concerning the specification of the basic behavioural rules of miniature robots giving rise to the emergence of a global collective behaviour. The two latters deal with the second part of the problematic, concerning the evaluation of the global collective behaviour resulting from the micro properties. We use a stochastic model by Markov's chains to evaluate the real conditions of the occurrence of the desired collective behaviour.

This paper focuses on the last stage of evaluation of the observability conditions of a desired behaviour (section 5). For the sake of the stochastic modelisation, we will briefly introduce the methodology *Cirta* (section 2) and describe the specification and the instanciation stages of this methodology (sections 3 and 4). The latters have been fully described in ([8]) and illustrated with collective behaviours of foraging ([7]) and tumour extraction ([6]). The different stages of *Cirta* will be illustrated in this paper using a collective foraging behaviour.

## 2  The methodology *Cirta*

Most authors rely on the concept of emergence without really defining what they mean by it or with unusable concepts like unpredictability. We propose an operational definition of emergence and which is inspired by Bunge, Forrest ([11]) and Lenay ([9]) definitions. We firstly suppose:

- we have a set of interacting entities or agents that can be expressed in a micro level theory or vocabulary;

- the interaction dynamics produces a global phenomenon that could be a regularity, a stable state or a process;
- this phenomenon is observed by an external observer or by the system itself, and expressed in a macro level theory.

A more detailed account of emergence can be found in [11] and [8].

In this work, we use this definition to distinguish the *micro* from the *macro* level and to emphasise interaction for producing a non-compositional phenomenon. It will correspond to the stabilisation process leading to the observation of new behavioural structures at the collective scale.

Giving these considerations, we build a methodological framework, named *Cirta*, to lead the emergence process of collective structures and behaviours. This methodology is organised in three stages:

1. The first stage is a specification phase where the desired task is linked to collective structures described as spatio-temporal patterns of interactions. It leads to derive the micro world properties (the minirobots and the environment) in order to observe a given collective behaviour.
2. The second stage corresponds to an instanciation phase where the minirobots and the environment properties are concretely specified. We also instanciate the associated patterns of perception of the environment properties.
3. The third stage is an evaluation phase where the conditions of observability of the supposed global behaviour are validated using a stochastic model.

The first stage could be viewed as a descending analysis of the global behaviour to derive individuals' one. Conversely, the last stage is an ascending synthesis of the micro world properties to effectively observe the supposed behaviour. The second stage is here to limit the actual occurrence evaluation to a coordinated sub-set of the possible patterns of interactions.

The three phases of *Cirta* will be illustrated in this paper with a collective foraging behaviour. The ecological niche of the robots contains a nest and potential food sources. The robots' group (called *collectivity*) have to explore the environment and to exploit food sources (Fig. 1). We suppose the minirobots have limited physical abilities and are endowed with an adaptive behaviour-based architecture. The latter is based on a set of sensory-motor loops and a motivation field to emerge interesting behaviours from each robot's interactions with its local environment ([8]).

In the next sections, we will briefly introduce the specification and the instanciation phases and will focus on the third phase of the *Cirta* methodology (see [8] for a more detailed description of the first stages).

## 3  The *Cirta* specification phase

The specification phase uses first order calculus formulas to describe collective behaviours as spatio-temporal patterns of interactions involving the dynamics of the minirobots and the dynamics of the environment.

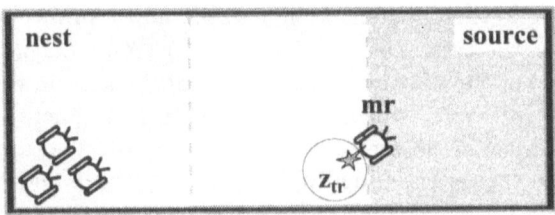

**Figure1.** Schema of the ecological niche

To briefly introduce the vocabulary necessary for describing collective actions and perception, we assume that the minirobots abilities are described in terms of stimulus-response pairs, and that each minirobot is only sensitive to the stimuli for which it has a response. We introduce the predicate $sensitive(mr, s, i, e)$ to state that the robot $mr$ is sensitive to the stimulus $s$ in the time interval $i$ and subspace $e$. In the same way, we will express that a robot $mr$ has a response $r$ for the stimulus $s$ by: $response(mr, s, r)$. This predicate does not mean that the robot will necessarily choose the response $r$ in presence of $s$, but only that this response is available.

To describe a task, the observer should be able to describe the environment with its properties including the minirobots ones. We introduce a set of propositions $p_i(c_1, \ldots, c_k)$ where $p_i$ is a property or a relationship and $c_i$ are variables or constants naming among others, the objects and the robots of the environment.

We introduce further the predicate $hold(prop, i)$ to state that the proposition $prop$ is true over the time interval $i$; the predicate $produce(prop, s, i, e)$ to state that $prop$ produces the stimulus $s$ in the time interval $i$ and subspace $e$, the predicate $effect_p(r, prop, i)$ to state that the response $r$ produces a property $prop$ over $i$. The formula $effect_s(r, s, i, e)$ states that the response can produce direct stimulus $s$ (direct robot-robot interaction); and the formula $effect(r, s, i, e)$ allows to talk directly about robot-robot interaction without making explicit the mediation by the environment.

### 3.1 Formalising collective actions

To achieve the foraging task, the *collectivity* needs to identify, localise and transport relevant objects of the environment, that can be formalised in terms of robots' interactions. In the following, we will only express the identification and localisation abilities (the complete formalisation could be found in [7] and [8]). We describe these collective abilities by introducing *tasks* and the operator $Can_c\, T$ to express that the *collectivity* can perform the task $T$.

We state in most cases (and in particular for the identification and localisation abilities) that a *collectivity* can perform a task if and only if at least an individual can perform the task:

$$Can_c\, T \overset{def}{=} \exists\, mr\ t.q.\ Can_{mr}\, T$$

Consequently, the following formalisation describe how a minirobot can perform a task. We consider the term $Identify(object, i, e)$ (resp. $Localise(object, i, e)$) describing the task of identifying (resp. localising in a near field) the object $object$ in the time interval $i$ within a subspace (e.g. a sphere) of detection $e$ associated to the object. A minirobot identifies the object in any time interval and subspace where the object's identifying property can be perceived by mean of a stimulus:

$$Can_{mr} Identify(object, i, e) \stackrel{def}{=} \exists \, p, s \; t.q. \; identify(p, object, i_p, e_p)$$
$$\wedge \; perceive(mr, p, s, i_s, e_s) \; \wedge \; i_s \subseteq (i \subseteq i_p) \; \wedge \; e_s \subseteq (e \subseteq e_p)$$

In addition to the ability of identifying the object, the robot should be in its neighbourhood and so should have a notion of proximity to localise the object. We must have a stimulus $s$ related to the proximity relation $p(mr, object)$. For example, $p$ can be expressed as a distance relationship $(distance(mr, object) < \delta)$:

$$Can_{mr} Localise(object, i, e) \stackrel{def}{=} Can_{mr} Identify(object, i, e)$$
$$\wedge \, \exists \, p, s \; t.q. \; identify(p, (mr, object), i_p, e_p)$$
$$\wedge \; perceive(mr, p, s, i_s, e_s) \; \wedge \; i_s \subseteq (i \subseteq i_p) \; \wedge \; e_s \subseteq (e \subseteq e_p)$$

The $identify(p, object, i_p, e_p)$ conditions ensure the properties to be locally non ambiguous. We shall now make explicit the perception of the environment properties and conditions.

## 3.2 Formalising perception of an environment condition

We specify the property of a minirobot $mr$ to be able to perceive a condition of the environment $prop$ by mean of a stimulus $s$ using the predicate $perceive$, which includes different ways to achieve the perception of a condition:

1. by detecting a stimulus that is directly tied to the property (e.g. by seeing directly a food source) or,
2. by an effect of the response to another stimulus produced by the same property (e.g. when a robot detects a pheromone trail that leads to the source, in which case $s$ is the perception of the trail) or,
3. by an effect of the response to anything which happens to produce this stimulus if and only if this property holds (e.g. following an obstacle as a response to a proximity stimulus can compute a curvature which is an internal stimulus linked to an object category),
4. it can further be produced when a team of robots reacting to different aspects of the property $prop$, creates collectively a state of the world that leads to the presence of the stimulus $s$ which is detected by the robot.

$$perceive(mr, prop, s, e, i) \stackrel{def}{=} \exists \, e_s, i_s \; t.q.$$
$$(produce(prop, s, e_s, i_s) \; \wedge \; sensitive(mr, s, e_{mr}, i_{mr})$$
$$\wedge \; e_{mr} \subseteq (e \subseteq e_s) \; \wedge \; i_{mr} \subseteq (i \subseteq i_s)) \tag{1}$$

$$\lor \ (\exists \ mr', s', e', i', r \ t.q. \ perceive(mr', prop, s', e', i')$$
$$\land \ response(mr', s', r) \ \land \ effect(r, s, e_s, i_s)$$
$$\land \ sensitive(mr, s, e_{mr}, i_{mr}) \ \land \ e_{mr} \subseteq (e \subseteq e_s) \ \land \ i_{mr} \subseteq (i \subseteq i_s)) \quad (2)$$
$$\lor \ (\exists \ prop', mr', s', e', i', r \ t.q. \ perceive(mr', prop', s', e', i')$$
$$\land \ response(mr', s', r) \ \land \ (hold(prop, e_p, i_p) \rightarrow effect(r, s, e_s, i_s))$$
$$\land \ sensitive(mr, s, e_{mr}, i_{mr}) \ \land \ e_{mr} \subseteq (e \subseteq e_s) \ \land \ i_{mr} \subseteq (i \subseteq i_s)) \quad (3)$$
$$\lor \ (\exists \ prop_j, prop'_j, mr_j, s_j, e_j, i_j, \ldots, j = 1 \ldots k \ t.q. \ (\wedge_{j=1}^{k} prop_j) \rightarrow prop$$
$$\land \ perceive(mr_j, prop_j, s_j, e_j, i_j) \ \land \ response(mr_j, s_j, r_j)$$
$$\land \ effect(r_j, prop'_j, e_{p'_j}, i_{p'_j}) \ \land \ produce(\wedge_{j=1}^{k} prop'_j, s, e_s, i_s)$$
$$\land \ sensitive(mr, s, e_{mr}, i_{mr}) \ \land \ e_{mr} \subseteq (e \subseteq e_s) \ \land \ i_{mr} \subseteq (i \subseteq i_s)) \quad (4)$$

Note that the above-mentioned responses can be produced by any robot including the same one and also by any coalition of minirobots. This large variety of interactions' possibilities leads to various patterns of activities in achieving collective actions.

## 4 The *Cirta* instanciation phase

The observation of coherent collective behaviours depends on well coordinated patterns of interactions. At this stage, the designer must concretely instanciate environment properties and minirobots stimuli, and specify which spatio-temporal interactions must happen in order for the task to be achieved. The idea is to select particular stimuli and strategies so to minimise the minirobots' abilities in terms of physical sensory-motor capacities and primitive behaviours. Technological constraints may also contribute to this selection but are not discussed here.

The analysis of the global behaviour according to the specification phase leads us to instanciate the collective abilities.

- For *the collective source identification*, we associate to the source a sonar signal as an identification property distinguishing in a non-ambiguous way a food source. This property is perceived directly by mean of the stimulus $s_s$ or also by seeing a track ($s_{tr}$, an infrared signal) produced by an other robot in response to the stimulus $s_s$ or even $s_{tr}$.
- In the same way, for *the collective nest identification*, we define the property of an other sonar signal (with a length wave different from the food source signal) that identifies a nest. This property is perceived directly by mean of the stimulus $s_n$.
- Finally, for *the collective nest and source localisation*, we suppose that these properties could be confused with the identification ones. In particular, we suppose that a suitable threshold of the sonar signal intensity related to a source determines the sufficient relation proximity of the robot to the food.

Note also that the perception of a track (an infrared signal emitted by a minirobot) has priority over direct exploration or source identification stimuli. The sign-stimuli[1] that could be perceived by a minirobot and the corresponding responses are summed up in the table (1).

**Table1.** Sign-stimuli and potential associated responses for a minirobot

| Sign-stimuli | Responses |
|:---:|:---|
| $S_n$ | flee from the nest, |
| $\left(S_{ex} \wedge \overline{S_{tr}}\right)$ | explore the environment, |
| $S_{tr}$ | pursue the track signal and propagate the track (emitting $S_{tr}$), |
| $S_{\overline{tr}}$ | try to track down a lost signal, |
| $\left(S_s \wedge \overline{S_{tr}}\right)$ | go towards the food source and create a track. |

## 5   The *Cirta* effective occurrence phase

The last phase of the methodology *Cirta* concerns the evaluation of the real conditions of occurrence of the supposed foraging behaviour. It uses a stochastic model by Markov's chains to study the dynamics of the *collectivity* and the effects of some crucial parameters such as the number of minirobots and the fields of influence[2] of relevant features (nest, sources ... ).

The environment dynamics being too complex and partially unknown to model, we have decided to model the dynamics of the minirobot's interactions (essentially of its perception). The relevant percepts are composed from the stimuli and called *sign-stimuli*. The dynamics is then expressed by the percepts dynamic vector $V_{mr}(t)$ (formula 5):

$$V_{mr}(t) = \mathcal{P}_{mr} \cdot V_{mr}(t-1) \qquad (5)$$

where $\mathcal{P}_{mr}$ is the transition matrix corresponding to the probabilities of transition of sign-stimuli perceptions. The transition matrix probabilities will capture generic features of the environment and the minirobot strategy.

Under these preliminary considerations, characterising the evolution of the minirobot amounts to specify the transition matrix. To do this, we will start from the instanciation phase where the stimuli to be perceived by a minirobot are described. According to these stimuli, we will then analyse the transitions of their perception and specify the corresponding graph of transitions (and from which the transition matrix is directly derived). Finally, we will evaluate the probabilities of transition to characterise the limit observed regularities.

---

[1] A *sign-stimulus* is a relevant combination of elementary stimuli.

[2] An *influence field* of an object, a property or a stimulus is defined as the spatio-temporal field of interactions involving it.

To simplify the probabilities expressions, we assume that the minirobots are homogeneous, i.e., they initially possess equivalent sensory-motor abilities. They could evolve differently according to their spatial position and sensory-motor past experiments. In the next sections, we will respectively describe the modelisation of the dynamics of a minirobot and of the dynamics of the tracks created in the environment.

## 5.1 The modelisation of the minirobot's dynamics

We assume that the dynamics of a minirobot could be characterised by a Markov's process which is discrete in time and homogeneous[3]. In our case study, the percepts dynamic vector is given in (6), standing for the probabilities of perception of each sign-stimulus in the minirobot's stimuli space. The minirobot's evolution is characterised by a set of trajectories in the space of the sign-stimuli[4] perceived by the minirobot.

$$V_{mr}(t) = \begin{bmatrix} P\left(S_n, t\right) \\ P\left((S_{ex} \wedge \overline{S_{tr}}), t\right) \\ P\left(S_{\overline{tr}}, t\right) \\ P\left(S_{tr}, t\right) \\ P\left((S_s \wedge \overline{S_{tr}}), t\right) \end{bmatrix} \tag{6}$$

The transition matrix $\mathcal{P}_{mr}$ defines the probabilities of transition of sign-stimuli perceptions (formula 7) and is deduced from the individual's abilities of source identification and localisation.

$$\mathcal{P}_{mr} = [p_{ij}] \quad \text{with} \quad p_{ij} = P(Ss_j; Ss_i) \ ; \ i, j = 1, \ldots 5 \tag{7}$$

In order to specify the matrix, we propose to represent the graph of the possible transitions between the sign-stimuli depending both on the environment and the minirobot's responses. The resulting graph from a recursive analysis of the perception of the related stimuli (according to the specification and instanciation phases of *Cirta*), is given in figure (2).

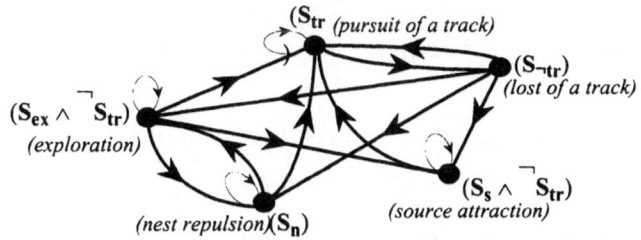

**Figure2.** The graph of perceptible sign-stimuli for a minirobot

---

[3] The transition probabilities are independent of time.

[4] These sign-stimuli are chosen each other exclusive to allow a stochastic modelisation.

Where for example, from an initial state of perception of the nest stimulus $(S_n)$, a minirobot could perceive in the next state the exploration area $(S_{ex})$, a track $(S_{tr})$ leading to a food source, or even the nest $(S_n)$.

From the graph of transitions (2), we derive the general form (8) of the matrix $\mathcal{P}_{mr}$ where the signs ($\bullet$) and ($\times$) denote respectively the transition probabilities $P(Ss_i; Ss_i)$ (diagonal elements) and $P(Ss_j; Ss_i)$, $i \neq j$:

$$\mathcal{P}_{mr} = [p_{ij} = P(Ss_j; Ss_i)] = \begin{bmatrix} \bullet & \times & \times & 0 & 0 \\ \times & \bullet & \times & 0 & 0 \\ 0 & 0 & 0 & \times & 0 \\ \times & \times & \times & \bullet & \times \\ 0 & \times & \times & 0 & \bullet \end{bmatrix} \qquad (8)$$

Note that the probability of transition $P\left(S_{\overline{tr}}; S_{\overline{tr}}\right)$ vanishes since the lost of a track is only memorised on one step.

We verify the strong connectivity and the non periodicity of the graph (2) associated to the transition matrix, which ensures the convergence of the minirobot behaviour to a limit behaviour given by the equation (9). In addition, because of these conditions, the Markov's process is regular and the limit vector is independent of the initial state of the minirobot:

$$V_{mr}(t = \infty) = \mathcal{P}_{mr}^{\infty} \cdot V_{mr}(t = 0) \qquad (9)$$

where the limit matrix $\mathcal{P}_{mr}^{\infty}$ has equal columns and the vector $V_{mr}(t = 0)$ corresponds to the initial state of the minirobot $mr$.

The transition matrix elements are deduced from the probability of perception of a stimulus formalised in the section (3.2). These probabilities include geometrical constraints as the overlapping of objects influence fields (nest, source and tracks), dynamic effects of responses and design choices as the number of robots and the distribution of sign-stimuli on the robots. A numerical evaluation of the probabilities of transition is then completed by considering the condition of stochastic matrix (the sum of each column's elements is equal to 1).

The matrix of transition of sign-stimuli perceptions for each minirobot $mr_i$ (index $i$) of the ecological niche is formulated in (10):

$$\mathcal{P}_{mr_i} = \begin{bmatrix} \frac{1/3}{2/3+(M-i)/27} & \frac{1/3}{1+(M-i)/27} & \frac{1/3}{1+(M-i)/27} & 0 & 0 \\ \frac{1/3}{2/3+(M-i)/27} & \frac{1/3}{1+(M-i)/27} & \frac{1/3}{1+(M-i)/27} & 0 & 0 \\ 0 & 0 & 0 & \frac{1-(M-i)/27}{1+8(M-i)/27} & 0 \\ \frac{(M-i)/27}{2/3+(M-i)/27} & \frac{(M-i)/27}{1+(M-i)/27} & \frac{(M-i)/27}{1+(M-i)/27} & \frac{9(M-i)/27}{1+8(M-i)/27} & \frac{(M-i)/27}{1/3+(M-i)/27} \\ 0 & \frac{1/3}{1+(M-i)/27} & \frac{1/3}{1+(M-i)/27} & 0 & \frac{1/3}{1/3+(M-i)/27} \end{bmatrix}$$
$$(10)$$

Finally, the study of the $kth$ powers of the transition matrix shows that $\mathcal{P}_{mr_i}^{\infty}$ is stabilised for an order of approximately ($k \approx 10$) for different robots indexes and total numbers of robots. The limit vector corresponding to the limit behaviour of the minirobot $mr_{i=1}$ is illustrated in figure (3) for different values of the total number of robots.

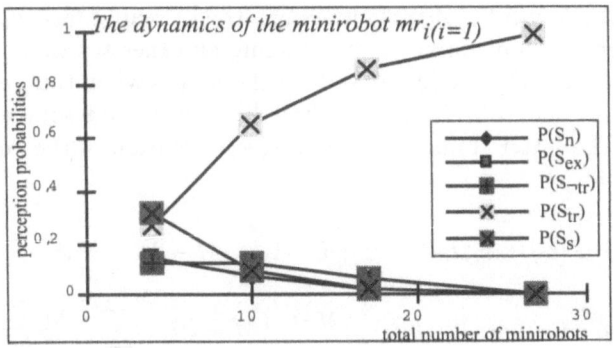

**Figure3.** The limit dynamic state vector of a minirobot versus the total number of robots ($M$)

This graphic shows that for small total numbers $M$ of robots, the probabilities of perception of a source or a track are equivalent and dominate the other probabilities. However, where increasing the total number of robots, the probability of direct perception of a source decreases and the probability of perception of a track is reinforced. Note that the total number of robots is here limited by the available space possibly covered with minirobots. These results let foretell that the global stable regularity observed at the collective scale is a behaviour of formation of minirobots' chains connecting the nest to food sources. This result will be confirmed when studying the dynamics of the created tracks in the environment.

## 5.2  The modelisation of the tracks' dynamics

In this section, we consider the dynamics of creation and propagation of tracks in the environment. A track is an infrared signal created by a minirobot when identifying a food source or when perceiving a track that leads to a source. The dynamics of the tracks' evolution is formulated by the dynamic state vector $V_{tr}(t)$ representing the probabilities of perception of a track by each minirobot $mr_i$ of the *collectivity* ($M$) at the time $t - i + 1$.

$$V_{tr}(t) = {}^T[P(mr_i, S_{tr}, t - i + 1), \ i = 1 \ldots M - 1]$$

We prove that this dynamic vector verify a fixed point equation of the form (11):

$$[V_{tr}(t)] = [\mathcal{A}] + [\mathcal{B}] \cdot [V_{tr}(t)] \tag{11}$$

where the matrix parameters $\mathcal{A}(mr_i, mr_{i+1})$, $i \leq M - 1$ and $\mathcal{B}(mr_i)$, $i \leq M - 2$ are deduced from the minirobots dynamics (section 5.1):

$$\begin{cases} \mathcal{A}(mr_i, mr_{i+1}) = P\left(mr_{i+1}, (S_s \wedge \overline{S_{tr}}), t - i\right) \cdot Pr\{z_{mr_i} \subseteq z_{track}\} \\ \mathcal{B}(mr_i) = Pr\{z_{mr_i} \subseteq z_{track}\} \end{cases}$$

expressing that the minirobot $mr_i$ is pursuing a track created by congener $mr_{i+1}$ that directly found a food source or by pursuing an other track. The boundary condition is given by $P(mr_M, S_{tr}, t - M + 1) = 0$ saying that at least one minirobot $mr_M$ of the *collectivity* has directly seen a food source and not a track related to the latter. Finally, the recursive resolution of the equation (11) gives:

$$
\begin{cases}
V_{tr}(t) = [v_{tr}(mr_i, t - i + 1), \; i = 1, \dots M - 1] \quad \text{with} \\
v_{tr}(mr_i, t - i + 1) = pp_s \cdot pp_{tr}^{M-i} \cdot (M - i)! \cdot \sum_{k=0}^{M-1-i} \frac{\left(\frac{1}{pp_{tr}}\right)^k}{k!}
\end{cases} \quad (12)
$$

where $pp_s$ and $pp_{tr}$ describe respectively the probability of perception of a source (given by the minirobot's dynamics) and the rate of the influence field of a track $z_{track}$.

Some components of the tracks dynamic vector are represented for the first minirobots $mr_i, i = M - 2, M - 1$ and the last ones $mr_j, j = 1, 2$ constituting a chain of tracks on the graph (4) versus the total number of the minirobots present in the ecological niche. The probabilities of perception of a track by

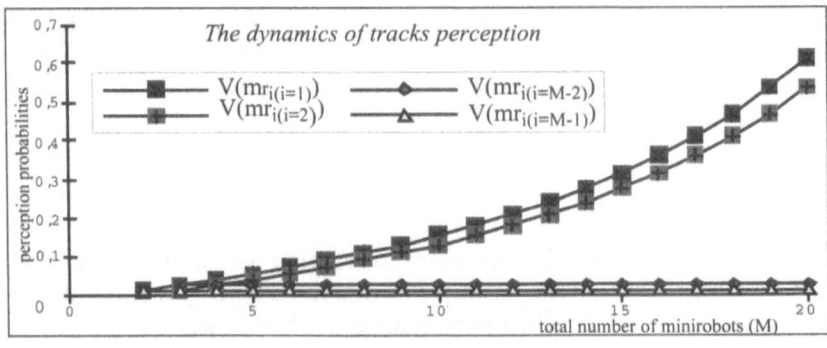

**Figure4.** Probabilities of a track perception versus the total number of minirobots

each minirobot of the *collectivity* (the whole vector components) are represented on the graphic (5) for different values of the total number of the minirobots ($M = 2, 5, 10, 15$ and $20$).

These graphics show that the probabilities of a track perception increase with the number of robots, that in turns create more tracks in the environment. However, these probabilities decrease with the index of the minirobots. The robots with high indices corresponding to the first elements of the group forming a chain of tracks, directly perceive a food source and so have less chance to pursue a track.

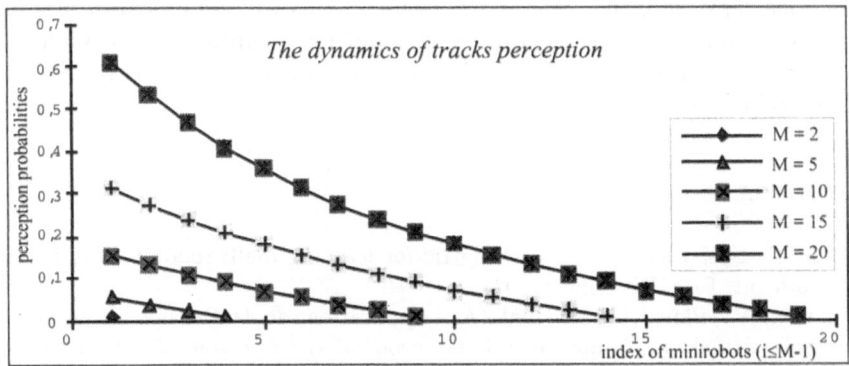

**Figure5.** Probabilities of a track perception versus the index of the minirobots

# 6 Conclusion

We have presented the methodology *Cirta* to design and evaluate emergent collective behaviours of robots' groups, where the *micro* world of individual properties and the *macro* world of global and collective structures are naturally articulated. It leads on the one hand to derive the micro world properties (the minirobots and the environment) in order to observe a given collective behaviour. On the other hand, it leads to an evaluation of the observability conditions of the desired collective behaviour. The methodology *Cirta* relies on a positive definition of emergence and on global models of the whole system to take into account the dynamics of robots-environment interactions in the genesis of emergent structures.

We have illustrated the three methodology stages using a simplified example of collective foraging and have emphasised the third stage of the effective occurrence evaluation of collective patterns of activities. This evaluation uses Markov's process to characterise the dynamics of the minirobots and the tracks created and propagated in the environment. According to our emergence definition, if such a process is stabilised it leads to structural and dynamic emergence.

We have shown that the limit regularity observed at the collective scale is the chain formation of minirobots connecting the nest to food sources. This result is essentially due to the recursive definition of perception that allows for large varieties of patterns of interactions for achieving an environment condition. In particular, *perception* does not appear as building any representation in the robot's head; but appears to be the result of the activity of the group which can, by interaction only, make present to one minirobot, conditions which are distant in space and time.

However, the stochastic model has neglected the possibility of formation of tree-like track structures when supposing that an influence field of a track could be occupied by only one minirobot. In the real world (according to experimental results), the range of infrared sensors in pursuit behaviour is sufficient to cover more that one robot. The stochastic modelisation methodology remains valid, but the consideration of non linear and cyclic chains will complicate the model during the evaluation phase of the conditional probabilities in the transition matrix. Finally, comparison of experimental results with the theoretical ones still have to be done.

# References

1. Balch, T.: Social entropy: A new metric for learning multi-robot teams. Proc. of the 10th Int. FLAIRS Conf. (1997)
2. Collinot, A., Ploix, L., Drogoul, A.: Application de la méthode Cassiopée à l'organisation d'une équipe de robots. Proc. of the 4th Journées Francophones d'IAD/SMA, Müller J.P., Quinqueton J. Coords. Hermes Edts (1996) 137–152
3. Ferber, J.: Coopération réactive et émergence. Intellectica **19** (1994) 19–52
4. Holland, J.H.: Hidden order: How adaptation builds complexity. Addison-Wesley (1995)
5. Kelly, K.: Out of control: The rise of neo-biological civilization. Addison-Wesley (1994)
6. Labbani, O., Müller, J.P., Bourjault, A.: Designing emergent collective behaviours: Application to a tumour extraction. Proc. of the IEEE/ASME Int. Conf. on Advanced Intelligent Mechatronics, Tokyo (1997)
7. Labbani, O., Müller, J.P., Bourjault, A.: Analyse en vue d'une conception de comportement collectifs émergents dans une colonie de robots. Proc. of the 5th Journées Francophones d'IAD/SMA, Quinqueton J. et al. Coords. Hermes Edts (1997) 343–357
8. Labbani-Igbida, O.: Contribution à une méthodologie de conception de comportements collectifs émergents dans une colonie de robots miniatures et autonomes. Phd thesis, University of Franche-Comté (1998)
9. Lenay, C.: Coopération et intentionalité. Proc. of the 4th Journées Francophones d'IAD/SMA, Müller J.P., Quinqueton J. Coords. Hermes Edts (1996) 265–272
10. Mataric, M.: Minimizing complexity in controlling a mobile robot population. Proc. of the IEEE Int. Conf. on Robotics and Automation, Nice (1992) 830–835
11. M.R. Jean (members of "Collectif" working group): Emergence et SMA. Proc. of the 5th Journées Francophones d'IAD/SMA, Quinqueton J. et al. Coords. Hermes Edts (1997) 323–341
12. Parunak, V.D.: "Go to the ant": Engineering principles from natural multi-agent systems. Multi-Agent Systems' Workshop, Kyoto (1996)
13. Smithers, T.: What the dynamics of adaptive behaviour and cognition might look like in agent-environment interaction systems. Proc. of Practice and Future of Autonomous Agents, Ticino (1995) 1–27
14. Steels, L.: Emergent functionality in robotic agents through on-line evolution. Proc. of the Alife IV Conference, MIT Press (1994)
15. Wooldridge, M.: Time, Knowledge, and Choice. in Intelligent Agents II, LNAI 1037, Springer Verlag, 1996

# Communication in Domains with Unreliable, Single-Channel, Low-Bandwidth Communication*

Peter Stone and Manuela Veloso

Computer Science Department
Carnegie Mellon University
Pittsburgh, PA 15213
{pstone,veloso}@cs.cmu.edu
http://www.cs.cmu.edu/{~pstone,~mmv}

**Abstract.** In most multiagent systems with communicating agents, the agents have the luxury of using reliable, multi-step negotiation protocols. They can do so primarily when communication is reliable and the cost of communication relative to other actions is small. Conversely, this paper considers multiagent environments with unreliable, high-cost communication. This paper presents techniques for dealing with the obstacles to inter-agent communication in such environments. A successful prototype system is fully implemented and tested in the simulated robotic soccer domain.

## 1 Introduction

In most multiagent systems with communicating agents, the agents have the luxury of using reliable, multi-step negotiation protocols (see [1] for instance). They can do so primarily when communication is reliable and the cost of communication relative to other actions is small. For example, in Cohen's convoy example [2], the communication time required to form and maintain a convoy of vehicles is insignificant compared to the time it takes the convoy to drive to its destination. Similarly, message passing among distributed information agents is typically very quick compared to the searches and services that they are performing. Thus, it makes sense for agents to initiate and confirm their coalition while guaranteeing that they will inform each other if they have trouble fulfilling their part of the joint action.

Conversely, this paper considers multiagent environments with unreliable, high-cost communication. For example, if there is only a single, low-bandwidth, unreliable communication channel for all the agents, and if the agents must

---
* This research is sponsored in part by the DARPA/RL Knowledge Based Planning and Scheduling Initiative under grant number F30602-95-1-0018. The views and conclusions contained in this document are those of the authors and should not be interpreted as representing the official policies or endorsements, either expressed or implied, of the U. S. Government.

sacrifice valuable resources in order to communicate, then although inter-agent communication may be beneficial, the agents' behaviors must not *depend* upon it.

One clear example of such an environment is the Soccer Server (version 3)—a widely used robotic soccer simulator—with a single, low-bandwidth, unreliable communication channel for all 22 agents and with high communication costs [8]. We use this domain for the research reported here. Another example domain is one that uses aural communication in crowded settings. Both people and robots using aural sensors ( [4]) must contend with multiple simultaneous audible streams. They also have a limit to the amount of sound they can process in a given amount of time, as well as to the range within which communication is possible. A third example of such an environment is arbitrarily expandable systems. If agents aren't aware of what other agents exist in the environment, then all agents must use a single universally-known communication channel, at least in order to initiate communication.

This paper presents techniques for dealing with the obstacles to inter-agent communication in such environments, particularly those with several *teams* of agents.

## 2 Team Member Architecture

Our new communication paradigm is situated within a team member architecture suitable for multiagent domains in which team members must act autonomously while working towards a common team goal. The team can synchronize ahead of time but while executing the task, communication is limited. Based on a standard agent architecture, our team member architecture allows agents to sense the environment, to reason about and select their actions, and to act in the real world. At team synchronization opportunities, the team also makes a *locker-room agreement* for use by all agents during periods of low communication. This section summarizes our team member architecture; it is described more fully in [11].

An agent keeps track of three different types of state: the *world state*, the *locker-room agreement*, and the *internal state*. The agent also has two different types of behaviors: *internal behaviors* and *external behaviors*.

The world state reflects the agent's conception of the real world, both via its sensors and via the predicted effects of its actions. It is updated as a result of processed sensory information. It may also be updated according to the predicted effects of the external behavior module's chosen actions. The world state is directly accessible to both internal and external behaviors.

The locker-room agreement is set by the team when it is able to privately synchronize. It defines the flexible team structure as presented below as well as inter-agent communication protocols. The locker-room agreement may change periodically when the team is able to re-synchronize; however, it generally remains unchanged. The locker-room agreement is accessible only to internal behaviors.

The internal state stores the agent's internal variables. It may reflect previous and current world states, possibly as specified by the locker-room agreement.

The internal behaviors update the agent's internal state based on its current internal state, the world state, and the team's locker-room agreement. The external behaviors reference the world and internal states, sending commands to the actuators. The actions affect the real world, thus altering the agent's future percepts. External behaviors consider only the world and internal states, without direct access to the locker-room agreement.

Internal and external behaviors are similar in structure, as they are both sets of condition/action pairs where conditions are logical expressions over the inputs and actions are themselves behaviors. In both cases, a behavior is a directed acyclic graph (DAG) of arbitrary depth. The leaves of the DAGs are the behavior types' respective outputs: internal state changes for internal behaviors and action primitives for external behaviors.

Some internal state variables need to be devoted to communication. When an agent hears a message, it interprets it and updates the world state to reflect any information transmitted by the message. It also stores the content of the message as a special variable `last-message`. Furthermore, based on the locker-room agreement, an internal behavior then updates the internal state. If the message requires a response, three variables in the internal state are manipulated by an internal behavior: `response`, `response-flag`, and `communicate-delay`. `response` is the actual response that should be given by the agent as determined in part by the locker-room agreement. All three of these variables are then referenced by an external behavior to determine when a response should be given. For example one condition-action pair of the top-level external behavior might be: `if (response-flag set and communicate-delay==0) then SAY(response)`.

Locker-room agreements can be used to eliminate or reduce the need for future communication, and they can also be used to increase communication reliability. For example, team members could agree upon a code number with which all messages should start in order to distinguish their messages from those of other teams in case other teams send similar messages on the single communication channel. They could also synchronize internal clocks if there is no globally accessible clock.

## 3  Communication Paradigm

The challenge for an agent to distinguish messages that are meant for it from those that are not is the first of five challenges that arise in the type of environment considered here. Second, since there is a single communication channel, agents must be prepared for active interference by hostile agents. A hostile agent could mimic messages it has previously heard at random times. Third, since the communication channel has low bandwidth, the team must prevent itself from all "talking at once." Many communication utterances call for responses from all team members. However, if all team members respond simultaneously, few of the responses will get through. Fourth, since communication is unreliable, agents must be robust to lost messages: their behaviors cannot depend upon receiving communications from a teammate. Fifth, teams must determine how to maximize the chances that they are using the same team strategy despite the facts that each is acting autonomously and that communication is unreliable.

| Communication Environment | Challenges |
|---|---|
| • many agents, teams | • message targeting/distinguishing |
| • single-channel | • robustness to active interference |
| • low-bandwidth | • multiple simultaneous responses |
| • unreliable | • robustness to lost messages |
| • high cost | • team coordination |

**Table 1.** The characteristics and challenges of the type of communication environment considered in this paper.

In order to meet these challenges, we propose that a team should use messages of the following form:

(<team-identifier> <unique-team-member-ID> <encoded-time-stamp> <time-stamped-team-strategy> <selected-internal-state> <target> <message-type> <message-data>)

Such a formulation assumes that the bandwidth allows for messages of several bytes in length to be transmitted in a reasonable amount of time. Some aural communication scenarios may need fewer, or condensed fields.

The contents of these fields are the product of the locker-room agreement. When forming the team, the agents must agree upon their team name (<team-identifier>) and a unique ID number for each member. For simplicity, the member IDs can be sequential numbers. These first two fields ensure that any teammate hearing the message knows precisely who uttered it. Teammates also agree ahead of time upon the security code used to create the field <encoded-time-stamp>. To coordinate, they agree upon a method for encoding and changing team strategies, and possibly upon positions of their internal states that should be communicated to help keep teammate information up to date. In addition, they must choose a set of acceptable message-types. The messages can use any syntactic and semantic codes (KQML [3] and KIF [5] for example). The only requirement is that the agents also agree on a mapping from message type to response requirements. Finally, the <target> field can be used to indicate the intended recipient(s) of the message. It could be intended for a single team member, for some subset of them, or for all team members.

The remainder of this section details how these particular message fields can be used to meet the challenges summarized in Table 1.

## 3.1  Message Targeting/Distinguishing

Agents can distinguish messages that are intended for them by checking the <team-identifier> and <target> fields. An agent $A$ listens to a message from a member of the same team that is targeted to $A$, to the entire team, or to some subset of the team that includes $A$[2]. All other messages may be ignored, or since all team members know the locker-room agreement, agents may use these messages to monitor their teammates' internal states.

## 3.2  Robustness to Active Interference

The only further difficulty related to an agent distinguishing which messages are intended for it arises in the presence of active interference. Consider a hostile agent $H$ which hears a message that is directed to $A$ at time $t$. $H$ has full access to the message since all agents use the same communication channel. Thus if $H$ remembers the message and sends an identical message at time $u$, agent $A$ will mistakenly believe that the message is from a teammate. Although the message was appropriate at time $t$, it may be obsolete at time $u$ and it could potentially confuse $A$ as $H$ intends.

This potential difficulty is avoided with the <encoded-time-stamp> field. Even a simple time stamp is likely to safeguard against interference since $H$ is not privy to the locker-room agreement: it does not necessarily know which field is the time stamp. However, if $H$ somehow discovers which field is the time stamp, it could alter the field based on the time elapsed between times $t$ and $u$. Indeed, if there is a globally accessible clock, $H$ would simply have to replace $t$ with $u$ in the message. However, the team can safeguard against such interference techniques by encoding the time-stamp using an injective function chosen as a part of the locker-room agreement. This function can use any of the other message fields as arguments in order to make decryption as difficult as possible. The only requirement is that a teammate receiving the message can invert the function to determine the time at which the message was sent. If the time at which it was sent is either too far in the past or in the future (according to the locker-room agreement), then the message can be safely ignored. Of course, it is theoretically possible for hostile agents to crack simple codes and alter the <encoded-time-stamp> field appropriately before sending a false message. However, the function can be made arbitrarily complex so that such a feat is intractable within the context of the domain. If secrecy is critical and computation unconstrained, a theoretically safe encryption scheme can be used. [3]

---

[2] The subsets could also be indicated by tokens if predetermined "units," or sub-formations, are formed.

[3] The degree of complexity necessary depends upon the number of messages that will be sent after the locker-room agreement. With few enough messages, a simple linear combination of the numerical message fields may suffice.

## 3.3   Multiple Simultaneous Responses

The next challenge to meet is that of messages that require responses from several teammates. However, not all messages are of this type. For example, a message meaning "where are you?" requires a response, while "look out behind you" does not. Therefore it is first necessary for agents to classify messages in terms of whether or not they require responses as a function of the <message-type> field. Since the low-bandwidth channel prevents multiple simultaneous responses, the agents must also reason about the number of intended recipients as indicated by the <target> field. Taking these two factors into account, there are six types of messages:

|  | Response requested | |
|---|---|---|
| Message Target | no | yes |
| Single agent | a1 | b1 |
| Whole team | a2 | b2 |
| Part of team | a3 | b3 |

When hearing any message, the agent should update its internal beliefs of the other agent's status as indicated by the <time-stamped-team-strategy> field. However, only when the message is intended for it should it consider the content of the message. Then it should use the following algorithm in response to the message:

1. If the message requires no response (cases a1-3), the agent simply updates its internal state.

2. If the message requires a response then set **response** to the appropriate response message, **response-flag** $= 1$ and

- if the agent was the only target (case b1), respond immediately: **communicate-delay** $= 0$;

- if the message is sent to more than one target (cases b2 and b3), set **communicate-delay** based on the difference between the <unique-team-member-ID> of the message sender and that of the receiver. Thus each teammate responds at a different time, leaving time for teammate messages to go through.

Then, if an internal behavior keeps decrementing **communicate-delay** as time passes, an external behavior can use the communication condition-action pair presented in Section 2: **if (response-flag set and communicate-delay==0) then SAY(response)**. Players can also set the **communicate-delay** variable in the event that they need to send multiple messages to the same agent in a short time. This communication paradigm allows agents to continue real-time acting while reasoning about the appropriate time to communicate.

## 3.4   Robustness to Lost Messages

In order to meet the challenge raised by unreliable communication leading to lost messages, agents must not depend on communication to act. Communication should be structured so that it helps agents update their world and internal states. But agents should not stop acting while waiting for communications from teammates. As brought up in [12], such a case could cause infinite looping if a critical teammate fails to respond for any reason. In the same way that agents continue acting while waiting for `communicate-delay` to expire, agents must do their best to maintain accurate world and internal states without help from teammates and continue acting while waiting for responses from teammates.

## 3.5 Team Coordination

Finally, team coordination is difficult to achieve in the face of the possibility that autonomous team members may not agree on the <time-stamped-team-strategy> or the mapping from teammates to roles within the team strategy. Again, there should be no disastrous results should team members temporarily adopt different strategies; however they should always do their best to stay coordinated. One method of coordination is via the locker-room agreement. Agents could agree on globally accessible environmental cues as triggers for switches in team strategy. Another method of coordination which complements this first approach is via the time stamp. When hearing a message from a teammate indicating that the team strategy is different from the agent's current idea of the team strategy, the agent adopts the more recent team strategy: if the received message's team strategy has a time-stamp that is more recent than that on the agent's current team strategy, it switches; otherwise it keeps the same team strategy and informs its teammate of the change. Thus changes in team strategy can quickly propagate through the team.

The <selected-internal-state> can also be used to facilitate team coordination by helping to keep the team members up-to-date regarding each other's status. Due to bandwidth constraints, it should not in general be an agent's entire internal state. However it might indicate the role that the agent is currently filling within the team strategy and any other particularly useful information as determined during the locker-room agreement.

## 3.6 Related Work

Most inter-agent communication models (as surveyed in [9]) assume reliable point-to-point message passing with negligible communication costs. In particular, KQML assumes point-to-point message passing, possibly with the aid of facilitator agents [3]. Nonetheless, KQML performatives could be used for the content portions of our proposed communication scheme. KQML does not address the problems raised by having a single, low-bandwidth communication channel.

With only a single team present, a situation similar to the one considered here is examined in [7]. In that case, like in ours, messages sent are only heard by agents within a circular region of the sender. Communication is used for cooperation and for knowledge sharing. Like in the examples presented in the soccer domain, agents attempt to update each other on their own internal states when communicating. However, the exploration task considered there is much simpler than soccer, particularly in that there are no opponents using the same communication channel and in that the nature of the task allows for simpler, less urgent communication.

Although communication in the presence of hostile agents is well studied in military contexts from the standpoint of encryption, the problem considered here is not the same. While any encryption scheme could be used for the message content, the work presented here assumes that the adversaries cannot decode the message. Even so, they can disrupt communication by mimicking past messages textually: presumably past message have some meaning to the team that uttered them. Our method of message coding based on a globally accessible clock circumvents this latter problem.

Even when communication time is insignificant compared to action execution, such as in a helicopter fighting domain, it can be risky for agents to absolutely rely on communication. As pointed out in [12], if the teammate with whom an agent is communicating gets shot down, the agent could be incapacitated if it requires a response from the teammate. This work also considers the cost of communication in terms of risking opponent eavesdropping and the benefits of communication in terms of shifting roles among team members. However, the problems raised by a single communication channel and the possibility of active interference are not considered, nor are the challenges raised when communication conflicts with real-time action.

A possible application of the method described here is to robots using audio communication. This type of communication is inherently single-channel and low-bandwidth. An example of such a system is the Robot Entertainment Systems which uses a tonal language [4]. Agents can communicate by emitting and recognizing a range of audible pitches. In such a system, the number of bits per message would have to be lowered, but the general techniques presented above still apply.

## 4   Implementation in the Robotic Soccer Domain

The soccer server [8] system used successfully at RoboCup-97 [6] during IJCAI-97 models a communication environment appropriate in a time-pressured, crowded environment. All 22 agents (11 on each team) use a single, unreliable communication channel. When one agent speaks, agents on both teams can hear the message immediately along with the (relative) direction from which it came. The speaker is not inherently known. Agents have a limited communication range, hearing only messages spoken from within a certain distance. They also have a limited communication capacity, hearing a maximum of 1 message every 200ms (actions are possible every 100ms, so if all other agents are speaking as fast as they can, only 1 of every 42 messages will be heard). Thus communication is extremely unreliable. Furthermore, on every 100ms action cycle, agents can either communicate or move in the world. Since the real-time nature of the domain requires quick and timely reactions, and since opponents hear all messages, communicating involves a significant cost.

---

- All 22 agents (including adversaries) on same channel
- Limited communication range and capacity
- No guarantee of sounds getting through
- Instantaneous communication
- High communication cost

---

**Table 2.** Characteristics of the Soccer Server communication model.

## 4.1 Our Communication Approach in the Soccer Server

In our team structure, players are organized into team formations with each player filling a unique role. However players can switch among roles and the entire team can change formations. Both formations and roles are defined as part of the locker-room agreement, and each player is given a unique ID number. It is a significant challenge for players to remain coordinated, both by all believing that they are using the same formation and by filling all the roles in the formation. Since agents are all completely autonomous, such coordination is not guaranteed. For more details on the implementation issues relating to this team structure, see [11].

As proposed in Section 3, all of our agent messages are of the form:

(CMUnited <Uniform-number> <Encoded-stamp> <Formation-number> <Formation-set-time> <Position-number> <target> <Message-type> [<Message-data>])

For example, player 8 might want to pass to player 6 but not know precisely where player 6 is at the moment. In this case, it could send the message (CMUnited 8 312 1 0 7 ----> 6 Where are you?). "CMUnited 8" is the sender's team and number; "312" is the <Encoded-stamp>, in this case an agreed upon linear combination of the current time, the formation number, and the sender's position number; "1 0" is the team formation player 8 is using followed by the time at which it started using it; "7" is player 8's current position; "----> 6" indicates that the message is for player 6; and "Where are you?" is a message type indicating that a particular response is requested: the recipient's coordinate location. In this case, there is no message data.

Upon hearing such a message, any teammate would update its internal state to indicate that player 8 is playing position 7. However only player 6 sets its **response** and **response-flag** internal state variables. In this case, since the target is a single player, the **communicate-delay** flag remains at 0. Were the message targeted towards the whole team or to a subset of the team, then **communicate-delay** would equal:

- IF (my number > sender number)

  ((my number − sender number − 1)\*2)\*`communicate-interval`
- ELSE (((sender number − my number − 1)\*2)+1)\*`communicate-interval`

where `communicate-interval` is the time between audible messages for a given agent (200ms in this case). Thus, assuming no further interference, player 8 would be able to hear responses from all targets.

Once player 6 is ready to respond, it might send back the message (`CMUnited 6 342 1 0 5 ----> all I'm at 4.1 -24.5`). Notice that player 6 is using the same team formation but playing a different position from player 8: position 5. Since this message doesn't require a response (as indicated by the "I'm at" message type), the message is accessible to the whole team ("`----> all`"): all teammates who hear the message update their world states to reflect the message data. In this case, player 6 is at coordinate position $(4.1, -24.5)$.

Notice that were player 8 not to receive a response from player 6 before passing, it could still pass to its best estimate of player 6's location: should the message fail to get through, no disaster would result. Such is the nature of most communication in this domain. Should there be a situation which absolutely requires that a message get through, the sending agent could repeat the message periodically until hearing confirmation from the recipient that the message has arrived. However, such a technique incurs high action costs and should be used sparingly.

Notice that in the two example messages above, both players are using the same team-formation. However, such is not always the case. Even if they use common environmental cues to trigger formation changes, one player might miss the cue. In order to combat such a case, players update the team formation if a teammate is using a different formation that was set a later time. For example, if player 6's message had begun "(`CMUnited 6 342 3 50 ...`" indicating that it had been using team formation 3 since time 50, an internal behavior in player 8 would have changed its internal state to indicate the new team strategy. Thus our team was able to remain coordinated even when changing formations.

Other examples of communication used in our implementation of simulated robotic soccer players include:

- Request/respond ball location
- Request/respond teammate location
- Inform pass destination
- Inform going to the ball
- Inform taking/leaving position

We found that the resulting updates of player world states and internal states greatly improved the performance of our team.

## 4.2 Results

Detailed empirical testing indicates that the implementation detailed above is successful in the challenging communication environment of the Soccer Server. In this section, we report results reflecting the cost of communication, the agents' robustness to active interference, their ability to handle messages that require responses from multiple team members, and their ability to maintain a coordinated team strategy.

To test the cost of communication, we played a team using no communication (team A) against a team identical to the first in all regards except that its members say random strings periodically (team B). Thus team B gained no benefit from communication, but its action rate was reduced by the interleaving of random statements. With an average of 18% of its actions taken by these random communications, team B suffered a significant degradation in performance, losing to team A by an average score of 3.54 to 1.08 over 50 games. Clearly, communication in this domain involves a significant cost.

Relying on communication protocols also involves the danger that an opponent could actively interfere by mimicking an agent's obsolete messages. However, our <Encoded-stamp> field guards against such an attempt. As a test, we played a communicating team (team C) against a team that periodically repeats past opponent messages (team D). Team C set the <Encoded-stamp> field to <Uniform-number> *(send-time + 37). Thus teammates could determine send-time by inverting the same calculation (known to all through the locker-room agreement). Messages received more than a second after the send-time were disregarded. The one-second leeway accounts for the fact that teammates may have slightly different notions of the current global time.

In our experiment, agents from team D sent a total of 73 false messages over the course of a 5-minute game. Not knowing team C's locker-room agreement, they were unable to adjust the <Encoded-stamp> field appropriately. The number of team C agents hearing a false message ranged from 0 to 11, averaging 3.6. In all cases, each of the team C agents hearing the false message correctly ignored it. Only one message truly from a team C player was incorrectly ignored by team C players, due to the sending agent's internal clock temporarily diverging from the correct value by more than a second. Although it didn't happen in the experiment, it is also theoretically possible that an agent from team D could mimic a message within a second of the time that it was originally sent, thus causing it to be indistinguishable from valid messages. However, in this case, the content of the message is presumably still appropriate and consequently unlikely to confuse team C.

Next we tested our proposed method of handling multiple simultaneous responses to a single message. Placing all 11 agents within hearing range, a single agent periodically sent a "where are you" message to the entire team and recorded the responses it received. In all cases, all 10 teammates heard the original message and responded. However, as shown in Table 3, the use of our proposed method dramatically increased the number of responses that got through to the sending agent. When the team used communicate-delay as specified in Section 4, message responses were staggered over the course of about 2.5 seconds, allowing most of the 10 responses to get through. When all agents responded at once (no delay), only one response (from a random teammate) was heard.

Finally, we tested the team's ability to maintain coordinated team strategies when changing formations via communication. One player was given the power

| | Number of Responses | | | Response Time (sec) | | |
|---|---|---|---|---|---|---|
| | Min | Max | Avg | Min | Max | Avg |
| No Delay | 1 | 1 | 1.0 | 0.0 | 0.0 | 0.0 |
| Delay | 6 | 9 | 8.1 | 0.0 | 2.6 | 0.9 |

**Table 3.** When the team uses communicate-delay as specified in Section 4, an average of 7.1 more responses get through than when not using it. Average response time remains under one second. Both experiments were performed 50 times.

to toggle the team's formation between a defensive and an offensive formation. Announcing the change only once, the rest of team had to either react to the original message, or get the news from another teammate via other communications. As described in Section 4, the <Formation-number> and <Formation-set-time> fields are used for this purpose. We ran two different experiments, each consisting of 50 formation changes. In the first, a midfielder made the changes, thus making it possible for most teammates to hear the original message. In the second experiment, fewer players heard the original message since it was sent by the goaltender from the far end of the field. Even so, the team was able to change formations in an average time of 3.4 seconds. Results are summarized in Table 4.

| | Entire Team Change Time (sec) | | | | Heard From |
|---|---|---|---|---|---|
| Decision-Maker | Min | Max | Avg | Var | Decision-Maker |
| Goaltender | 0.0 | 23.8 | 3.4 | 17.8 | 46.6% |
| Midfielder | 0.0 | 7.9 | 1.3 | 2.8 | 80.6% |

**Table 4.** The time it takes for the entire team to change team strategies when a single agent makes the decision. Even when the decision-making agent is at the edge of the field (goaltender) so that fewer than half of teammates can hear the single message indicating the switch, the team is completely coordinated after an average of 3.4 seconds.

In addition to the above controlled experiments, we used our communication method in the CMUnited simulator team [10] that competed in RoboCup-97. In a field of 29 teams, CMUnited made it to the semi-finals, indicating that the overall team construction, of which this communication model was a significant part, was successful.

# 5 Conclusion

In domains with low-bandwidth, single-channel, unreliable communication, several issues arise that need not be considered in most multiagent domains. We have presented a communication paradigm which successfully addresses these challenges. Having fully implemented it in the robotic soccer domain, we have tested the paradigm empirically both in a controlled setting and in competition against several previously unseen opponents. Using this paradigm, the CMUnited-97 simulator team made it to the semi-finals of RoboCup-97.

# References

1. Mihai Barbuceanu and Mark S. Fox. Cool: A language for describing coordination in multi agent systems. In *Proceedings of the First International Conference on Multi-Agent Systems (ICMAS-95)*, pages 17–24, Menlo Park, California, June 1995. AAAI Press.
2. Philip R. Cohen, Hector J. Levesque, and Ira Smith. On team formation, 1997.
3. Tim Finin, Don McKay, Rich Fritzson, and Robin McEntire. Kqml: An information and knowledge exchange protocol. In Kazuhiro Fuchi and Toshio Yokoi, editors, *Knowledge Building and Knowledge Sharing*. Ohmsha and IOS Press, 1994.
4. Masahiro Fujita and Koji Kageyama. An open architecture for robot entertainment. In *Proceedings of the First International Conference on Autonomous Agents*, pages 435–442, Marina del Rey, CA, February 1997.
5. M. R. Genesereth and R. E. Fikes. Knowledge interchange format, version 3.0 reference manual. Technical Report Logic-92-1, Computer Science Department, Stanford University, 1992.
6. Hiroaki Kitano, Yasuo Kuniyoshi, Itsuki Noda, Minoru Asada, Hitoshi Matsubara, and Eiichi Osawa. RoboCup: A challenge problem for AI. *AI Magazine*, 18(1):73–85, Spring 1997.
7. Dario Maio and Stefano Rizzi. Unsupervised multi-agent exploration of structured environments. In *Proceedings of the First International Conference on Multi-Agent Systems (ICMAS-95)*, pages 269–275, Menlo Park, California, June 1995. AAAI Press.
8. Itsuki Noda and Hitoshi Matsubara. Soccer server and researches on multi-agent systems. In *Proceedings of the IROS-96 Workshop on RoboCup*, November 1996.
9. Peter Stone and Manuela Veloso. Multiagent systems: A survey from a machine learning perspective. Technical Report CMU-CS-97-193, Computer Science Department, Carnegie Mellon University, Pittsburgh, PA, December 1997.
10. Peter Stone and Manuela Veloso. The CMUnited-97 simulator team. In Hiroaki Kitano, editor, *RoboCup-97: The First Robot World Cup Soccer Games and Conferences*. Springer Verlag, Berlin, 1998. In Press.
11. Peter Stone and Manuela Veloso. Task decomposition and dynamic role assignment for real-time strategic teamwork. In *5th International Workshop on Agent Theories, Architectures, and Languages*, July 1998.
12. Milind Tambe. Teamwork in real-world, dynamic environments. In *Proceedings of the Second International Conference on Multi-Agent Systems (ICMAS-96)*, Menlo Park, California, December 1996. AAAI Press.

# MARCH : A Flexible Multi-agent Architecture, Applied to Autonomous Robots Playing Football

Sébastien Rocher and Dominique Duhaut

Laboratoire de Robotique de Paris, Université PARIS 6, UVSQ, CNRS, France

**Abstract.** Our research is linked with the general pattern of computing architecture, dedicated to multi-robots applications. The design of a
robot football team needs to realize the different standard layers of an
autonomous mobile robot (perception, decision, action). It also includes
specific Artificial Intelligence and multi-agent systems problematics.
Our study is based on real robots (than we call physical agents). The
physical aspect is essential in this work. Indeed, we have to take account
of the physical constraints associated to our application. So, the behaviors that we use must be adapted to the technical capacity (the dexterity)
of the robots (the precision and the complexity of their actions). Moreover, an experimental platform is developed to validate the theoretical
physical agent behaviors. This platform is called MICROB.
In this paper, we propose a flexible Multi-agent ARCHitecture (MARCH)
adapted to multi-mobile-robots applications, for the football game purpose. In this architecture, we will propose the notions of phases and
roles to select appropriate behavior for each agent. We describe also, the
experimental platform and the physical agents.

## 1 Introduction

In our study, we try to define a general model for robotic applications involving several mobile robots. We take an interest in dynamical environment and
applications which requires cooperation.

The realization of a team of robots playing football, is well adapted to this
study [5]. Indeed, the goal of football can be defined like this: "to score more goals
than the opponent". It involves that all robots of the application are manœuvring
in the same dynamic real environment. Moreover, a team of robots will have to
wrestle with elements (the robots of the other team) which will try to wreck the
goal. So, the behavior and multi robots cooperation study are essential. For those
reasons this research area is well attended in Distributed Artificial Intelligence
researches [6], and several methods were proposed applied to simulation and to
real robots [1, 9, 7].

The objective of our work is to realize a real robot team, using physical agents.
The difficulties of this work arise from the physical constraints, the reliability
and the performance of the different elements of the global platform [2]. The
principal originality of our approach is to include the notion of phase, which
separate different kind of behavior by the goal of the application and the roles,
which support the strategies associated to reach this goal.

In the following section we introduce the flexible Multi-agent ARCHitecture (MARCH). We will particularly develop the game phases and the role notions, as well as the agent model. In the third section, we will study how MARCH can be applied to the football application. In the fourth one, we will present the experimental platform used. We will notably describe the robots architecture and the vision system. The implementation and the experiments will be discussed in the fifth section. And finally, we will conclude the work exposed in the sixth section.

## 2 MARCH: A Flexible Multi-agent ARCHitecture

The objective of our research was to define and realize an flexible architecture permitting to realize a multi-physical-agent applications. We can break a general multi-physical-agent application down into several entities : the application phases, the roles of each agent, the physical-agents, the behaviors and the actions. Those different entities represent the different abstraction level used to define an application.

The characteristic and the interest of this breakdown of an application is its flexibility. Indeed, this breakdown of an application is general and does not depend on any specific application. So, we will be able to systemize this breakdown of an application to define a very large number of multi-physical-agent applications.

So, we present now the different entities of our application.

### 2.1 The application phases

Robotic missions and multi-agent applications can be defined like a sequential series of phases. A phase of an application is a period during which, the general goal to achieve by the set of agents of the application (or the goals to achieve by each agent) is particular. So, when the application phase changes, the tasks to realize by each agent of the application will change, and then, the behaviors of each ones will have to be modified to be well adapted to the new goals. To use a sequential series of phases to define a mission, allows use to describe applications, not only with one main goal to achieve, but like a series of goals (or sub-goals) to achieve.

The football game is a good example to show the utility of this description. Indeed, the main goal to achieve in a football game is to score more goals than the opponent. But the football game obeys rules. Then, we can define different phases of the game like the kick-off, to take a corner, the penalty kick, etc. The set of these phases represents the football game. So, the change of phase depends on a set of rules associated to the application. We will develop in more details this subject in the next section.

In this part, we will particularly define the notions of phase and the notion of role. Then, we will present our multi-agent architecture.

## 2.2 Roles and strategy

Independently of the game phases, multi-agent applications brings us to study the cooperation and the organization problematics. Indeed, the efficiency of a multi-agent system is based on the definition of a collective strategy to achieve a specific goal. In this work the realization of a strategy leads to define a role for each physical-agent of our application. Of course, the roles will not be the same for each physical-agent (they will rather be complementary).

A role can be define as a set of available behaviors. Then, for each role and for each game phase, we will associate, for each physical-agent of the application, a specific behavior. Of course the strategy will have to be modified on line, during the execution of the mission.

## 2.3 The flexible architecture

On the basis of these comments, we propose to integrate the different entities of our application into a multi-agent architecture which is applicable to a large number of physical multi-agent applications (*fig 1*).

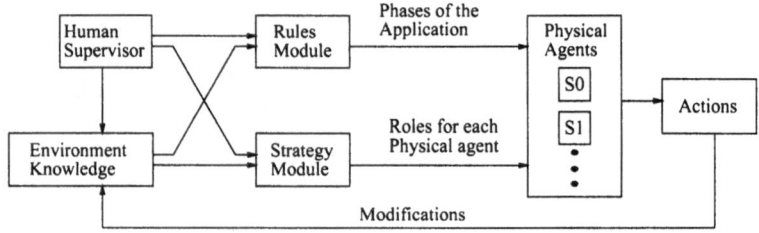

**Fig. 1.** *The flexible Multi-agent ARCHitecture: MARCH*

The architecture is composed of five main modules:

- the **human supervisor** who can define rules or strategy of the application,
- the **environmental knowledge** which is composed of the sensors data and of the static environmental knowledge,
- the **rules module** which defines the application phase from the rule supervisor and the environmental knowledge,
- the **strategy module** which defines the roles for each agent of the application from the strategy supervisor and from the environmental knowledge.
- the **set of agents**.

## 2.4 The agent model

The behavior of each robot depends on the game phase (defined by the game rules), on its role (defined by the strategy) and on partials informations of the environment (*fig 2*).

We can modelize the agents:

$A = \{A_i\}, i \in [0, N]$: set of N agents of the system.

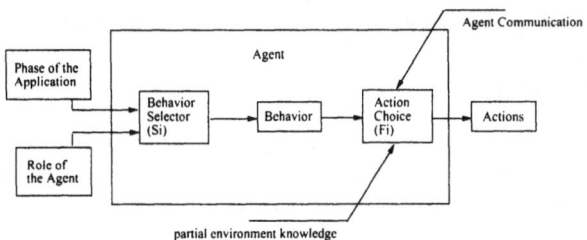

**Fig. 2.** *The agent model*

$B_i = \{B_{i_0}, ..., B_{i_k}\}$: set of k behaviors of the agent i.
$R_i = \{R_{i_0}, ..., R_{i_l}\}, R_{i_j} \in \mathcal{N}$: set of roles.
$P_i = \{P_{i_0}, ..., P_{i_m}\}, P_{i_j} \in \mathcal{N}$: set of phases.
$I_i = \{I_{i_0}, ..., I_{i_n}\}$: set of environment information known by the agent i.
$M_i = \{M_{i_0}, ..., M_{i_p}\}$: set of internal agent memory.
$C_i = \{C_{i_0}, ..., C_{i_q}\}$: set of available actions.
The function $s_i$ used to select the behavior is:
$$s_i : R_i \times P_i \to B_i.$$
The function $f_i$ used to choose the action is:
$$f_i : B_i \times I_i \times M_i \to C_i.$$

## 3 Application of MARCH to a football player robots application

The football game is a good test-bed for mobile robots application architectures. So we wished to apply MARCH to this application. In this part, we will study how to use the architecture proposed, to the football application. We will describe, notably, the functionalities of each modules of our system.

### 3.1 The environmental knowledge

In our system, there is two kinds of environmental knowledge :

- the **static knowledge** : the contour of the field, the kickoff position and the penalty position.
- the **dynamic knowledge** : as it is explained before, the video system computes the position and the orientation of the robots and the position and the velocity vector of the ball.

### 3.2 The rules module

The rules module act for a large part under the influence of the human supervisor. Indeed, he decides to stop or to start the game, etc... The refereeing is done by a human, so it is impossible to systematize all this module. The different phases we implemented to define the football game application are :

- the **kick-off** phase (for the opponent and for our team),
- the **free kick** phase (for the opponent and for our team),

- the **penalty kick** phase (for the opponent and for our team),
- the **stop** phase,
- the **main game** phase.

In certain cases, the module of rules allows to change automatically game phases : from the kick-off phase or the penalty kick phase, or the free kick phase to the main game phase. The rules can be written like this :

- *If ((GamePhase = KickOffPhase) and (BallVelocity > K))*
  *Then GamePhase = PrincipalGamePhase*
- *If ((GamePhase = OpponentFreeKickPhase) and (BallVelocity > K))*
  *Then GamePhase = PrincipalGamePhase*
- *If (GamePhase = OurTeamFreeKickPhase)*
  *Then GamePhase = PrincipalGamePhase*
- *...*

### 3.3 The strategy module

The goal of the strategy module is to define the role of each physical agent of the team. This role can be static or dynamic, according to the environmental knowledge. Two main roles are defined :

- the **goal keeper** role,
- the **field player** role.

At most, one agent can have the goal-keeper role during a game, and the field player role is common to several agents. To organize the team, we decided to characterize the field player roles of a agent, affecting him a specific space on the game field. So, the role of the field players physical agents will be differentiated by their activity field space. So, we will be able to have a defender player, a middle field player, a forward player, etc. Then, we will have subclasses of the main roles. Moreover, specific roles can be added to the principal ones. These are :

- the **penalty kicker** role,
- the **Kick-Off kicker** role.

### 3.4 The agent behaviors

A behaviors is selected according to the role of an agent and the game phase. The selected behavior will define the action to execute by the physical agent according to the environmental knowledge and the inter-agent communication. The behaviors will have to obey to the game rules. So, we expose here the type of behavior used for the football physical-agent application and the table representing the selection function (*fig 3*). The different implemented behaviors are :

- the **specific goal keeper** behavior,
- the **specific field player** behavior,
- the **keep stopped** behavior,
- the **kickoff player** behavior,
- the **penalty kick player** behavior.

| Game Phase | | Kick Off | | Free Kick | |
|---|---|---|---|---|---|
| Role | | Opponent | Our Team | Opponent | Our Team |
| Goal Keeper Specific Role | Penalty Player | Specific Goal Keeper Behavior | Specific Goal Keeper Behavior | Specific Goal Keeper Behavior | Specific Goal Keeper Behavior |
| | Kickoff Player | Specific Goal Keeper Behavior | Play the Kickoff Behavior | Specific Goal Keeper Behavior | Specific Goal Keeper Behavior |
| | None | Specific Goal Keeper Behavior | Specific Goal Keeper Behavior | Specific Goal Keeper Behavior | Specific Goal Keeper Behavior |
| Field Player Specific Role | Penalty Player | Keep Stopped Behavior | Keep Stopped Behavior | Keep Stopped Behavior | Specific Field Player Behavior |
| | Kickoff Player | Keep Stopped Behavior | Play the Kickoff Behavior | Keep Stopped Behavior | Specific Field Player Behavior |
| | None | Keep Stopped Behavior | Keep Stopped Behavior | Keep Stopped Behavior | Specific Field Player Behavior |

| Game Phase | | Penalty Kick | | Stop | Principal Game |
|---|---|---|---|---|---|
| Role | | Opponent | Our Team | | |
| Goal Keeper Specific Role | Penalty Player | Keep Stopped Behavior | Play the Penalty-Kick Behavior | Keep Stopped Behavior | Specific Goal Keeper Behavior |
| | Kickoff Player | Keep stopped Behavior | Specific Goal Keeper Behavior | Keep Stopped Behavior | Specific Goal Keeper Behavior |
| | None | Keep Stopped Behavior | Specific Goal Keeper Behavior | Keep Stopped Behavior | Specific Goal Keeper Behavior |
| Field Player Specific Role | Penalty Player | Keep Stopped Behavior | Play the Penalty-Kick Behavior | Keep Stopped Behavior | Specific Field Player Behavior |
| | Kickoff Player | Keep Stopped Behavior | Keep Stopped Behavior | Keep Stopped Behavior | Specific Field Player Behavior |
| | None | Keep Stopped Behavior | Keep Stopped Behavior | Keep Stopped Behavior | Specific Field Player Behavior |

**Fig. 3.** *Representation of the behavior selection function*

## 4 The football experimental platform

The platform used for our application is based on the experimental platform realized for the MICROB project [4]. The objective of this project is to study the collective organization behaviors applied to a robot society. To implement the robot football application, we kept the flexible architecture of the MICROB platform. The various modules of our system were simply adapted to our specific application, and we included our multi-agent architecture to determine and select the behaviors of each robot. The platform flexible architecture is composed with five different modules (*fig 4*). Those modules are :

- a **vision system** which process data from a video camera placed above the playing field,
- a **behavior module** which defines the adapted behaviors and the appropriated basic action to execute, for each robot,
- a **software control module** which allows to transform a basic action into an actuator order,

- a **communication module** which beams the order wished for each robot using an hertzian transmission,
- a **low level control** which transmits actuators torques.

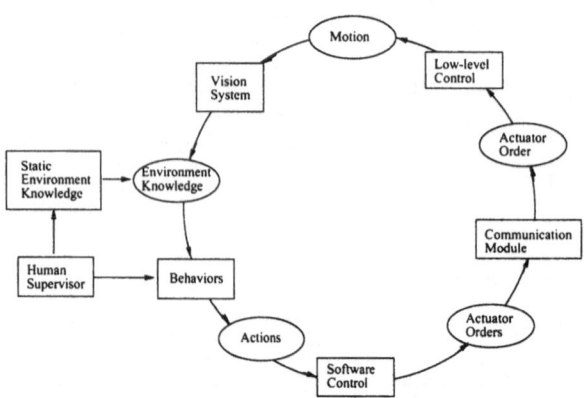

**Fig. 4.** *The flexible platform architecture*

The playing field of the application has the dimension of a ping-pong table and each team owns five players. The size of each robot can't be higher than 10cm×18cm for the small size league. Of course, the system has to realize a real-time processing. First, we will briefly present the robots and the vision system used. The cooperation and behavior research will be developed on the next section.

### 4.1 The robots architecture

The robots carried out have two independent wheels on the middle of their structure. So, they have two degrees of freedom in the plane. The physical structure of the robots is divided in four levels *(fig 5)* :

- the **actuators level** : the actuators used have got a tachometric feed-back. On this level, there is also the batteries.
- the **electronic level** : it changes the actuator orders in tension.
- the **CPU level** : the CPU only has to decode the frame coming from the herzian receiver of the robot.
- the **herzian receiver level** : it receives order from the order processing computer.

### 4.2 The vision system

The vision system takes a large part in the global system efficiency [8]. It has to extract pertinent data frame the video acquisition. Those pertinent data are :

- the **position** and the **orientation** for each robot,

**Fig. 5.** *One of our small size robot*

- the **position**, the **orientation** and the **velocity** of the ball.

Moreover, because of the real-time constraints, the acquisition and the processing of a video frame had to take less than $1/10^{th}$ of a second. The acquisition card used works in black and white . The image processing realizes a contour extraction to detect the objects placed in the playing field. To identify the robots, we put various marks on the top of each ones. The direction the robots are determined using the knowledge of their contour. For the ball, the direction and the velocity are deduced using the last detected positions. The precision of the detection is approximately 1cm.

# 5 Implementation and experiments

In the previous section, we presented the behavior selection function. In this section, we will develop the implementation of the behaviors, and we will study the function used to choose the action.

## 5.1 The actions

We distinguish two kinds of actions : the **high-level** actions and the **low-level** ones. The high-level actions define goals to realize (replacement, ball kicking, avoidance, move for a position with an orientation). These actions will be broke down to executable actions. These low-level actions must have the characteristic to be robust. They are :

- the **forward gear** ,
- the **reverse gear**,
- a **position and orientation low-level servoing**.

The third action is the most interresting one. To implement this action, we used a control system well adapted to our robots configuration, and to our problematic. This control system is detailled in a paper [3]. It allows a position

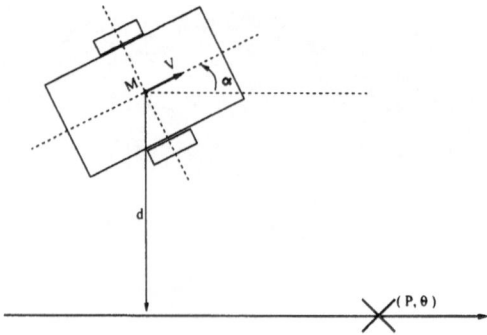

**Fig. 6.** *General description*

and orientation servoing using a constant linear velocity of the position M of the robot (*fig 6*).

Indeed, the velocity of each wheels is :

- the **right wheel** : $V_r = V - \Delta V$,
- the **left wheel** : $V_l = V + \Delta V$,

The velocity V is the velocity at the point M of the robot. It is a constant : $V = \frac{V_r + V_l}{2}$. We can determine $\Delta V$ using the formula :
$\Delta V = V(\frac{d}{K_1} + \frac{\alpha}{K_2}), with K_1 and K_2 constants.$
In our experiments, $K_1 = 60$ and $K_2 = 1.5$.

## 5.2 The behaviors description

In the section three, we enumerated five behaviors : the specific goal keeper behavior, the specific field player behavior, the keep stopped behavior, the kickoff player behavior and the penalty kick player behavior. The three last behaviors are obvious to implement. So, we will only explain the goal keeper behavior and the specific field player one.

**The goal keeper behavior** The goalkeeper behavior is very simple : during the game, we only place the goal keeper on a line segment, between the ball and the goal

**The specific field player behavior** The field player behavior is much more interesting. The specificity of our work is to use different activity spaces for each different field player behavior. The algorithm of the behavior can be resumed in a pseudo-code like this :
*if the ball is not in the activity space of the agent then*
    **Replacement.**
*else*
    *if the agent is not well placed to kick the ball then*

*Replacement.*
*else*
    *the agent **communicate to other agents the evaluation of its position to kick the ball.***
    *if the agent is the best placed then*
        *the agent **kick the ball.***
    *else*
        ***Replacement.***
    *end if*
    *end if*
*end if*

The idea is to have stackables activity spaces in the robot team. So, each time, several robots can kick the ball to attack or to defend. Finaly, only one robot will kick the ball (the one which is the best placed), and the other ones execute a replacement.

## 5.3 The experimentation

Two configurations were experimented to validate our system : one game with a five players team and one game with a two players team.

**The five players game** For the game with five players, we defined the following roles :

- one goal keeper,
- one defender player,
- one left middle field player,
- one right middle field player,
- one forward player.

To define the activity space assigned to each robot, we decided to divide the game field into 9 parts (*fig 7*). Then, each field player has a specific activity place.

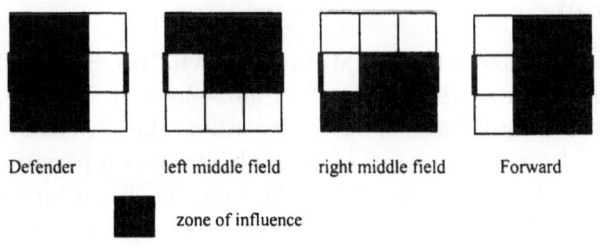

Defender    left middle field    right middle field    Forward

zone of influence

**Fig. 7.** *Space activity for each field player robot*

We have to define too, the replacement position. In our experiment, the replacement position only depends on ball position(*fig 8*).

During all the game with five players, the agent role always stole the same.

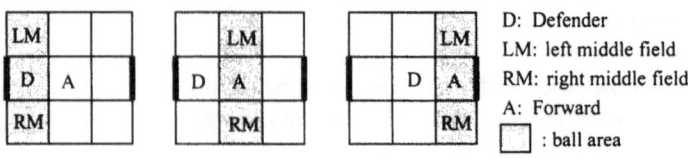

**Fig. 8.** *Replacement positions*

**The two players game** The game with two players is very different from the one with five players. Indeed it is much more difficult to kick the ball with two players than with five. So, we used this configuration :

– one defender player/goal keeper player,
– one forward player,

For the game with only two agents, we decided to give the maximum of freedom to our robots. So the activity space is the same for the two robots and this space represents all the game field. But the behaviors are not exactly the same for the two robots. Indeed, the replacement position is different (*fig 9*).

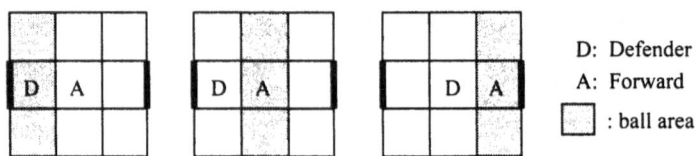

**Fig. 9.** *Replacement positions*

This configuration is particularly interesting, because it allows us to test the function of role selection. Indeed, the role of one of the robots is dynamic. It plays defender or goal keeper according to the environmental knowledge. If the ball is in our camp and if the direction of the ball is toward our goal, then the strategy module imposes to one agent the goal keeper role.

# 6  Conclusion

In this paper, we presented a flexible Multi-agent ARCHitecture : MARCH. This architecture is based on a break down of a multi-physical-agent application into several entities : the application phases, the roles, the physical-agents, the be-haviors and the actions. The phases of an applications is a period during which, each physical agent have to realize a specific task. Then, an application can be define as a sequential series of phases. Moreover, the notion of role allows us to represent and to study cooperative behaviors. The architecture proposed does not depend on a particular application. So we propose a flexible system, which allows an on-line change of the strategy (and so, the role of the physical-agents) according to the situation.

We experimented this architeture for a football game application. So, we presented the MICROB platform, which implement five autonomous mobile robots.

Several configurations were experimented to validate our sytem : one game with five robots, and one game with two robots. Very simple actions have been used for this application to deal with the physical constraint of our system (inaccuracies).

The experiment shows us the efficiency of our system to describe and control a general multi-physical-agent application. Now, the experimental platform is being improved to allow us to use more complex actions.

# References

1. M. Asada, E. Uchibe, S. Noda, S. Tawaratsumida, K. Hosoda, "Coordination of multiple behaviors acquired by vision-based reinforcement learning," *Proceedings of the International Conference on Intelligent Robots and systems (IROS)*, pp 917-924, 1994.
2. M. Asada, M. Mataric, Y. Kuniyoshi, D. Duhaut, A. Drogoul, P. Stone, H. Asama, H. Kitano, "The RoboCup Physical agent challenge: Phase-I," *The First International Workshop on RoboCup, in Conjunction with IJCAI-97*, pp 51-56, 1997.
3. S. Delaplace, P. Blazevic, J.G. Fontaine, N. Pons, J. Rabit, "Trajectory tracking for mobile robot" *S.G. Tzafestas (ed.) Robotic Systems* pp 313-320,1992 Kluwer Academic publishers.
4. A. Drogoul, D. Duhaut, "MICROB: Projet commun de LRP-LAFORIA 1995 de robotique collective," *Internal report*, 1995.
5. H. Kitano, M. Veloso, H. Matsubara, M. Tambe, S. Coradeschi, I. Noda, P. Stone, E. Osawa, M. Asada, "The robocup synthetic agent challenge 97," *The First International Workshop on RoboCup, in Conjunction with IJCAI-97*, pp 45-50, 1997.
6. H. Kitano, Y. Kuniyoshi, I. Noda, M. Asada, H. Matsubara, E. Osawa, "Robocup: a challenge problem for AI," *AI magazine Vol. 18 N°1*, pp 73-85, 1997.
7. J.L. de la Rosa, A. Oller, J. Veh, J. Puyol, "Soccer team based on agent-oriented programming" *Journal of Robotics and Autonomous System Vol. 21, N° 2*, pp 167-176, 1997.
8. R. Sargent, B. Bailey, C. Witty, A. Wright, "The importance of fast vision in winning the first micro-robot world cup soccer tournament" *Journal of Robotics and Autonomous System Vol. 21, N° 2*, pp 139-147, 1997.
9. P. Stone, M. Veloso, "Using decision tree confidence factors for multiagent control," *The First International Workshop on RoboCup, in Conjunction with IJCAI-97*, pp 31-36, 1997.

# Decision Trees and Rule Induction in Simulated Soccer Agents

Ioan Alfred Letia, Marius Joldos, Calin Cenan, Diana Zaiu, and Alina Andreica

Technical University, Department of Computer Science, Cluj-Napoca, Romania
letia@cs.utcluj.ro
http://cs-gw.utcluj.ro/people/letia/

**Abstract.** Low-level individual behavior and higher-level collaborative behavior are presented in a two-level layered architecture for simulated robotic soccer. Shooting to the goal and kicking in a passing situation are achieved by a neural network trained for various positions of the attacker and the goal-keeper or teammate, respectively. As collaboration with teammates passing is considered in various configurations of a group of four attackers and four defenders. Learning the decision of a player: (i) keep ball, (ii) pass ball to closest teammate, (iii) pass ball to medium distance teammate, (iv) pass ball to farthest teammate are studied. The algorithms used to learn this decision-making are OC1 and ITI, for decision tree, and CN2 and RIPPER for rule induction. These results are useful in constructing a decision-maker for a player.

*Keywords*: Robotic soccer, Learning low-level behavior, Learning higher-level behavior, Collaborative behavior with decision trees, Rule induction

## 1 Introduction

Robotic soccer is indeed a very reach problem domain to study and experiment concepts and solutions for multi-agent systems. Within the simulated soccer play set that we are developing, we have selected in this paper the problem of skill learning by an individual player and collaborative learning of a group of players [7]. The individual skills considered are shooting to the goal in the presence of a goal-keeper and receiving the ball from a pass of a teammate. Collaboration is represented by the decision on the act of one player within a group of players in the presence of opponents: keep the ball or pass to one of several teammates.

Learning low-level individual behavior has been given considerable attention [10, 16], but it still constitutes a challenge for an actual play set. Coordination in robotic soccer has been approached through decision-tree structures [4], [5], [6] and general teamwork formalisms, like joint intentions and shared plans [14]. Learning basic collaborative behavior has been approached mainly by neural networks [8] and decision trees [11], [13], [12]. We, however, use the neural network model to learn a low-level function and not the decision when to accelerate and when to stay or if a shoot is a better decision than a pass, as is done in these papers.

For a better understanding of the collaborative and adversarial aspects in simulated robotic soccer, we have chosen in this study to get some insight in the learning of passing within a group of players, by using various decision tree (DT) and rule induction (RI) algorithms, namely OC1 [9], ITI [15], CN2 [1] and RIPPER [3].

## 2  Soccer Team Architecture

Our current simulated robotic soccer team shown in Figure 1 has a model of the world for each agent and two layers: (i) *reactive level*, with basic operators, including neural network ones; (ii) *decision layer*, with decision trees and rules, obtained by induction.

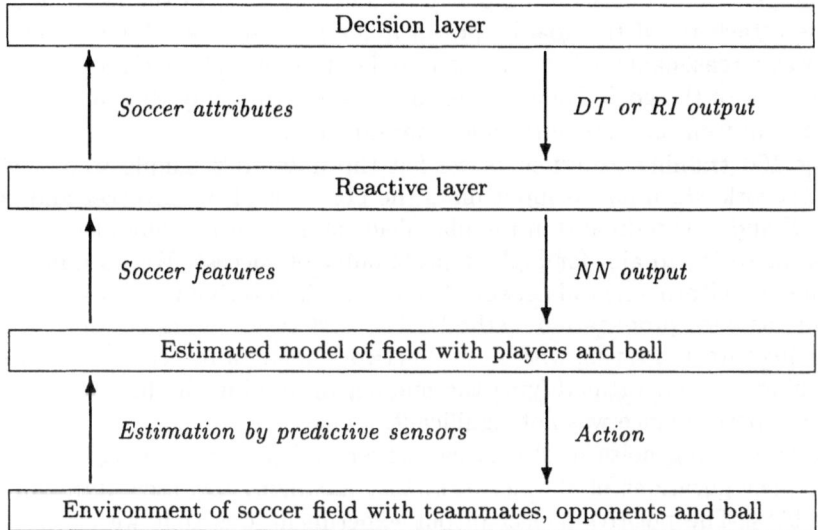

**Fig. 1.** Overview of layered architecture

## 3  Low-Level Behavior

As low-level behavior we illustrate our approach for shooting to the goal and kicking the ball in a pass to a teammate.

## 3.1 Shooting to the Goal

One of the first capabilities needed for our soccer agent was considered that of shooting to goal. We try to learn an agent to kick a fixed ball and score a goal. To further simplify this first experiment we also take into account only a fixed goal-keeper and no other opponents to block the space between the shooter and the goal. Instead of a procedure to shoot to goal, as in [11], what we want to acquire in this experiment is a function for the low-level behavior of the player in which to calculate the direction of the kick operation. We use the same neural network model but we hope to achieve with this learning mechanism a better generalization over the possible positions of the players in the field.

In the first experiment, we consider two static agents: an attacker which has a stationary ball in front of him and a goal-keeper. In contrast with the experiments presented in [11], where the shooter has to choose the moment to accelerate and the direction is predefined, in our case the shooter's task is to choose a point of the goal at which to aim, to kick the ball in the corresponding direction and score a goal.

In the first phase of the experiment, to gather training data, we randomly kick the ball at 10 different points of the goals and we decide which is the best. Trying to generalize our shooting policy to different areas of the field we move both the attacker and the goal-keeper in different positions. Moving the goal-keeper was a reasonable policy for this role, i.e. to move left or right on a line 5 units in front of the goal. The attacker moves were randomly chosen in an area of 30 units in front and 50 units wide from the goal.

After 100 training experiments we fed the data to a simple feed-forward neural network which has as input units the positions of the attacker and goal-keeper (X and Y coordinates in the play field) and as output units the point of the goal at which to aim for highest probability of success. We do quite a lot experiments to learn a neural network from these data trying to stick at a simple algorithm as back-propagation methods. The best results were obtained with a 0.5 learning rate and trying with several network configurations we noticed that adding hidden layers or modifying the number of units in the hidden layer the increase in performance was not significant.

Due to existing noise in the soccer server and in our training data, with different configurations of the network, the maximum accuracy obtained was around 95%. The positive aspect of our experiment was that we observe the possibility to generalize the behavior over the a large space of possible player positions. After training on these 100 patterns, we tested the resulted network on 1000 situations and we obtained a pretty good accuracy around 75%.

## 3.2 Kicking the Ball in a Pass to Teammate

Another skill needed for our soccer agent is the ability to pass. In our next experiment we tried to learn a player to kick in such a way that a second player, the receiver of the pass, be able to intercept the ball. This second experiment was inspired by [8] but instead of deciding between pass or shoot to goal we want to obtain another function for the low-level behavior of the agent to calculate the acceleration of the pass to a teammate situated at a given distance.

To learn passing, we place two static players on the play field: a passer (player 2) and a receiver (player 1). The task of the passer is to choose an acceleration and to kick the ball to the receiver. The receiver of the pass will try to observe the ball and if he sees the ball he must accelerate trough ball and kick, in which case the trial is recorded as positive. The coach will end any operation if the ball stops moving. To gather training data we have chosen a random policy to shoot with different forces and established which was the best acceleration for a given distance. To generalize over distances we moved the receiver of the pass at several distances in a domain between 5 to 35 units.

After 100 training experiments we fed the data to a simple feed-forward neural network which has as input units the relative coordinates of the receiver and as output units the acceleration of the pass which produced the highest probability of interception. Using back-propagation methods with the 0.5 learning rate, which produced acceptable results in the previous experiments, and with two hidden layers we learned these function. Afterwards we tested the learned network and we noticed an accuracy of around 85%.

## 4 Decision Trees for Collaborative Behavior

The scenario for collaborative behavior we investigated, where the player with the ball has to decide whether to pass the ball to one of his teammates taking into account the presence of opponents and not just the position of his teammates in that particular situation. The classes of actions considered in all experiments discussed here are: C2 → keep the ball; C3 → pass ball to closest teammate; C4 → pass ball to medium distance situated teammate in group; C5 → pass ball to farthest teammate in group.

The attributes of a passing position are described using the following abbreviations: d → distance; a → angle; t → teammate; o → opponent. For example, the meaning of dt3 is distance to closest teammate and the meaning of ao2 is angle for the next closest opponent.

On the decision level of the soccer player, to model a collaborative behavior with other agents from the environment, the agent must be able to pass (ability obtained on the low-level) and to decide to which teammate to pass. To choose from several receivers of the pass we used inductive methods to learn decision trees or rules for the soccer agent.

The goal of the learning process is to use these attributes to calculate the destination of a pass with greater success rate. For training, we placed a passer and randomly around him five opponents and three potential receivers. After observing a number of played matches we consider that these are reasonable assumptions for a scenario in which the soccer agent has to pass within a 30 units circle around him. The passer task was to choose a teammate and kick the ball using appropriate power and direction. If the receiver was able to intercept the ball he told the coach to record the instance as successful and if one of the opponents was able to intercept the ball he told that the instance was unsuccessful. We used a set of 300 samples for training with several algorithms and 100 situations for testing.

## 4.1 Oblique Decision Tree

Oblique decision trees [9] are trees in which each node may contain a (linear) multivariate test on data attributes. OC1 constructs oblique decision trees from examples. It can also construct standard axis-parallel trees, which contain tests of just one attribute in each node. Oblique decision trees are therefore a natural extension of the well-known axis-parallel trees.

A decision tree produced by OC1 from the training set is shown below, with some hyper-planes removed for lack of space.

```
Root Hyperplane: Left = [52,111,39,26], Right = [38,4,7,2]
2.399403 dt3 + 14.427605 at3 + -0.939261 dt4 + -0.811557 at4
 + -0.249277 dt5 + 0.264201 at5 + -0.569576 do1 + -0.506959 ao1
 + -3.061336 do2 + 9.561970 ao2 + 0.260523 do3 + -0.017392 ao3
 + -0.454480 do4 + 7.211802 ao4 + -0.871826 do5 + 2.219645 ao5
 + 50.277100 = 0

1 Hyperplane: Left = [15,4,3,10], Right = [37,107,36,16]
-0.312868 dt3 + -2.363012 at3 + 0.696809 dt4 + 0.337263 at4
 + 0.664301 dt5 + -8.972887 at5 + 0.930517 do1 + -0.767421 ao1
 + 0.406375 do2 + 13.015273 ao2 + -0.507178 do3 + 1.279513 ao3
 + -0.572978 do4 + 0.264851 ao4 + -0.047032 do5 + -0.120215 ao5
 + 4.564990 = 0

11 Hyperplane: Left = [5,1,1,10], Right = [10,3,2,0]
1.000000 at4 + 0.701508 = 0

1r Hyperplane: Left = [21,51,4,1], Right = [16,56,32,15]
-2.733725 dt3 + 3.476715 at3 + -0.192411 dt4 + -6.995778 at4
 + -1.768927 dt5 + 20.668503 at5 + -0.064354 do1 + -9.392509 ao1
 + 1.850529 do2 + 2.886914 ao2 + -1.737212 do3 + 1.328368 ao3
 + 0.150862 do4 + 0.977132 ao4 + 2.254603 do5 + -4.680733 ao5
 + 14.564844 = 0
```

OC1 can construct axis parallel trees as well. In this case only axis parallel splits at each node of the DT are performed. This is one way of using approximate axis parallel methods such as C4.5 and CART. Here is an axis parallel tree obtained from the same training set.

```
Root Hyperplane: Left = [68,51,13,7], Right = [19,60,32,20]
1.000000 do5 + -34.473801 = 0
```

```
1 Hyperplane: Left = [15,0,0,2], Right = [53,51,13,5]
1.000000 do5 + -20.431992 = 0

lr Hyperplane: Left = [49,36,10,2], Right = [4,15,3,3]
1.000000 do2 + -21.106709 = 0

lrl Hyperplane: Left = [23,28,5,2], Right = [26,8,5,0]
1.000000 at3 + -0.441722 = 0

lrll Hyperplane: Left = [7,0,1,0], Right = [16,28,4,2]
1.000000 ao5 + 1.216483 = 0
```

The oblique tree calculates the decision using all the continuous attributes which describe the global state of the environment. The axis parallel tree obtained is smaller than the previous one and uses mainly angle attributes and attributes related to opponents which seems to have a greater importance in the decision process. Testing the oblique decision tree we observed a more accurate behavior than for the axis parallel tree. We also notice similarities in the behavior of the player agent in the two cases when the oblique decision tree and the ITI decision tree are used.

## 4.2 Incremental Tree Induction

Incremental Tree Induction (ITI) [15] builds a decision tree in incremental or batch mode. In incremental mode, ITI incorporates each instance into the tree, and restructures the tree as needed so that it becomes the same tree that one would have gotten with the batch algorithm. Incremental induction is usually much less expensive than rebuilding the tree from scratch. Such a behavior is especially needed if we are interested in an on-line tracking of agents actions to further improve the ability to pass during a real game. In this approach the decision tree with which we have to start the program must be good enough but also flexible to permit the soccer agent to accommodate with different opponents in an adversarial environment.

For ITI a variety of training modes are possible but we use here the batch mode because it runs more quickly than incremental mode and builds the same tree. We have an initial set of 300 examples and no current tree so we build a decision tree as quickly as possible in batch mode. The virtual pruning is also used and the accuracy of the algorithm obtained is different. Virtual pruning is done according to the minimum description length principle.

Part of the DT obtained for passing is shown in Figure 2.

In addition to incremental induction, the ability to restructure a tree makes it possible in many cases to travel through tree-space inexpensively. The program includes direct metric tree induction (DMTI) (non-incremental) in which it tries various tests at a node, and evaluates the quality of each resulting tree.

We have tested also three variants of DTMI, each using a different direct metric: minimum description length, expected number of tests for classification

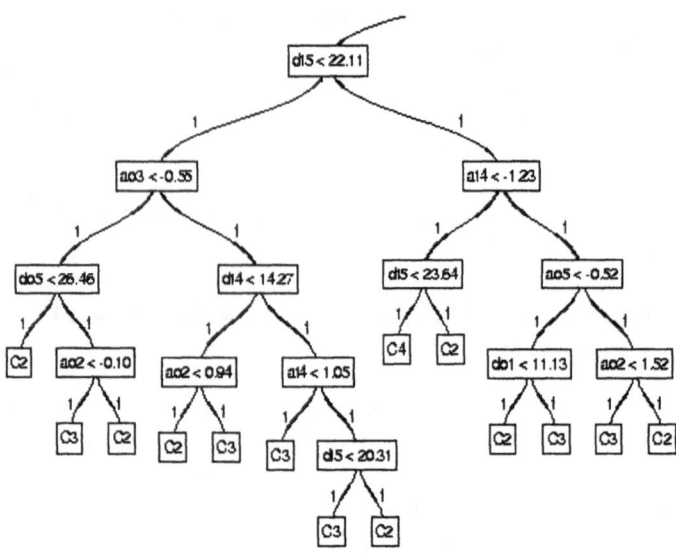

**Fig. 2.** ITI generated decision tree for passing

and number of leaves. The accuracy obtained on the test set was, in order: 34%, 37% and 41%. In the test set we had 100 examples and all the algorithms were applied to four classification tasks C2 (keep the ball), C3 (pass ball to closest teammate), C4 (pass ball to medium distance situated teammate) and C (5pass ball to farthest teammate).

It is observable from the obtained decision tree that the angle attributes are more important in the decisional process. We assume that this behavior is reasonable in choosing a player with a greater probability of receiving the ball. The choice in the first test is based on distance to the farthest player but on the next levels of the induction tree tests which check the angle to players became predominant. We also notice that the opponents distribution has greater influence to the decision than the teammates distribution.

## 5 Rule Induction for Collaborative Behavior

### 5.1 CN2

CN2 [1] is a rule-induction program which takes a set of examples (vectors of attribute-values), and generates a set of rules for classifying them. It also allows one to evaluate the accuracy of a set of rules (in terms of a set of pre-classified examples). The aim of the algorithm is to find rules whereby the class attribute of an example may be inferred from the non-class attributes.

The list of numbers at the end of each rule indicates the number of training examples covered by that rule, divided into classes. The precise significance of the counts depends on whether the rules are ordered or unordered. The counts associated with each rule in a set of unordered rules (this is our case), comprise all the examples covered by that rule including those which may be covered by other rules as well. This applies equally to the default rule, whose count therefore comprises the whole example set.

The evaluation module takes a set of rules and a pre-classified set of examples, and compares the classification given by the rules with the given class-values. The evaluation is of a set of rules taken as a whole. The results are displayed in the form of a matrix. The entry in row $i$, column $j$ of the matrix is the number of examples classified by the rules as class $j$ which were really of class $i$. Fractional values may arise if the examples include "Unknown" or "Don't Care" values.

```
IF     do1 < 6.20
  AND ao1 < 1.02
  AND do2 < 12.56
THEN  class = C2   [18 0 0 0]


IF     at5 < 1.29
  AND do5 < 20.17
  AND ao5 < 0.86
THEN  class = C2   [12 0 0 0]


IF     dt3 > 8.43
  AND do1 < 8.30
  AND ao1 > 1.09
  AND ao3 > -1.04
THEN  class = C2   [12 0 0 0]


IF     at4 > -1.12
  AND do1 > 8.21
  AND do4 > 14.05
  AND do5 < 21.83
THEN  class = C2   [5 0 0 0]


IF     at3 > -1.43
  AND at4 > -1.42
  AND do1 > 11.22
  AND ao2 < -0.39
THEN  class = C3   [0 13 0 0]


IF     dt4 < 7.94
  AND ao3 < 0.87
  AND ao4 < 0.80
THEN  class = C4   [0 0 3 0]


IF     ao2 < -1.46
  AND do3 < 25.86
```

```
    AND ao4 > -0.53
    THEN  class = C5  [0 0 0 3]
```

For example, given the above definitions, the first rule is interpreted as follows: the passer should keep the ball if the distance to the closest opponent is less than 6.2 units and angle he sees the closest opponent is less than 1.2 radians and also the distance to the next closer opponent is less than 12.56 unit. There were 18 examples covered by this rule in the training set.

## 5.2  RIPPER

RIPPER[3] is a program for inducing classification rules from a set of pre-classified examples. The user provides a set of examples, each of them labeled with the appropriate class. RIPPER will then look at the examples and find a set of rules that will predict the class for new examples.

RIPPER has several advantages over other learning techniques. First, RIPPER's hypothesis is expressed as a set of if-then rules. These rules are relatively easy for people to understand; if the ultimate goal is to gain insight into the data, then RIPPER is probably a better choice than a neural network learning method, or even a decision tree induction system. Second, RIPPER is asymptotically faster than other competitive rule learning algorithms; this means that it will be much faster on large datasets. Third, RIPPER allows the user to specify constraints on the format of the learned if-then rules. If there is some prior knowledge about the concept to be learned, then these constraints can often lead to more accurate hypotheses. Fourth, RIPPER allows attributes to be either nominal, continuous, or "set-valued". Recent versions of RIPPER also support bag-valued attributes. The basic strategy used by RIPPER is to find an initial model and then iteratively improve it using heuristic techniques and is more efficient on large noise datasets.

On our training set the rules produced by RIPPER were the ones below.

```
C4 10 5 IF ao5 <= 0.058756 do4 >= 29.5466 .
C2 36 15 IF do2 <= 13.6015 at3 >= 0.54042 .
C2 10 0 IF do5 <= 27.2947 dt4 >= 18.7883 dt3 <= 15.2643 .
C3 116 107 IF .
```

# 6   Experimental Results and Discussion

A programmable coach client was designed in order to perform the experiments. The coach client had the ability to communicate with the player clients via the soccer server. The synchronization for starting the experiments was achieved by communication (a specific message from coach to all players). The tasks for all players were set using the coach, who also set and checked the starting positions. During all experiments and at the end of each of them, the coach collected relevant data. The decision on ending the experiment was made by the coach, based on the conditions stated by the experimenters. The coach had two operation modes: an interactive mode, used to test the ideas for new experiments, and a batch mode, where an experiment was repeated for a large number of times, using randomly chosen positions. These positions were subject to constraints, depending on the goal of the experiment, in order to obtain a realistic simulation of real game situations.

The player clients are implemented in C++ (GNU). They make use of a modified version of Itsuki Noda's library, libsclient-0.03, for implementing the sensor and communication functions. The experiments were carried on Sun Ultra machines, running Solaris 2.5.1, and on PC's running Linux (installed from Slackware 3.0 and Red Hat 5.0 distribution kits). The soccer server used was initially Soccer Server version 3.28 and later version 4.03.

We used 300 training examples and 100 test examples for learning the passing decision and obtained the results shown in Table 1. ITI has been used with no pruning and pruning (+p), DMTI with no pruning and pruning, OC1 with no pruning and pruning (+p), OC1 with axis parallel (OC1p) without pruning and with pruning (+p), and OC1 with the CART perturbation algorithm.

**Table 1.** Decision tree induction results

|  | ITI | ITI +p | DMTI | DMTI +p | OC1 | OC1 +p | OC1p | OC1p +p | OC1 +CART |
|---|---|---|---|---|---|---|---|---|---|
| Leaves | 108 | 43 | 86 | 30 | 73 | 7 | 14 | 6 | 51 |
| Depth | 10 | 6 | 8 | 6 | 11 | 12 | 84 | 5 | 50 |
| Training accuracy | 100 | 77 | 100 | 66 | 94 | 69 | 94 | 55 | 46 |
| Test accuracy | 40 | 39 | 37 | 38 | 47 | 48 | 42 | 46 | 43 |

Evaluating the set of rules generated by CN2 on the test set we obtained the following results, which are presented in Table 2, the correct state versus the predicted one, followed by the accuracy for the state (with an overall accuracy of 63.5%).

**Table 2.** Rule induction results

|  | C2 | C3 | C4 | C5 | Accuracy |
|---|---|---|---|---|---|
| C2 | 31 | 15 | 0 | 0 | 67.4% |
| C3 | 8 | 75 | 1 | 0 | 89.3% |
| C4 | 4 | 27 | 11 | 0 | 26.2% |
| C5 | 4 | 10 | 1 | 5 | 25.0% |

We can conclude from here that the number of misclassified examples is small, so the right decision is taken in the majority of cases.

We notice in Table 3 that CN2 generates more rules than RIPPER, but the accuracy on the training and test sets was considerably better.

**Table 3.** Rule induction results

|  | RIPPER | CN2 |
|---|---|---|
| Rules | 3 | 53 |
| Training accuracy | 58 | 77 |
| Test accuracy | 46 | 63 |

The problem with RIPPER was that the program will always guess that a certain class, with the largest number of occurences, is a default class and this behavior was the cause for a not so high accuracy. CN2, in addition of the rules induction, has tried to fuzzyficate the parameters to cope with this qualitative tasks. In the soccer domain, where parameters are all continuous and a certain role is played by the noise in all aspects, we obtained with CN2 the best results, in terms of greatest accuracy and fewer parameters in rules.

In [11] the high-level decision that has been learned was whether or not to make a pass. In our work we considered several decisions that the agent has to make at one point in time: keep the ball, pass ball to closest teammate, pass ball to medium distance situated teammate, pass ball to farthest teammate. This task seems to be rather complex for an inductive learner. Although some of the statistics obtained do not appear very successful, given the complexity of the task and alternatives offered by combinations of decision-makers, we regard the results as quite encouraging.

## 7 Conclusions

The first experiments carried out towards setting up a simulated robotic soccer team have been encouraging. We have been able to learn the players to perform some low-level and higher-level actions reasonably well. However, these steps are only the initial ones towards more performative players.

We need to extend the experiments to cover more situations from actual soccer games, by collecting a larger and more significant set of examples. We also need a deeper understanding of the ways each of the methods represent the knowledge of the game. It seems that more advanced methods of combining prediction for decisions might be necessary, given the fact that different algorithms arrive at different solutions. Bagging and boosting should be worth trying in future work. Another further line of research will be to see what advantage we can gain by using qualitative models to guide inductive learning [2].

Of course pursuing this line of decision making will need to take into account the dynamics of the play, which was ignored for the moment. We also did not consider variations of play which is very important for a team in order to increase its chance of success and to make tracking by the opponent team more difficult. However, in each state of the game the agent finds itself at one specific moment of time, it has to make decisions based on the available attributes. Learning these decisions is quite a complex problem, but we favour the rules for their visibility and ease of inclusion in other decision mechanisms.

## Acknowledgements

In our RoboCup team environment we used the library provided by Itsuki Noda. We are grateful to the anonymous reviewers for their critical comments that helped us improve the paper considerably.

## References

1. Peter Clark and Robin Boswell. Rule induction with CN2: some recent improvements. In Yves Kodratoff, editor, *Proceedings of the Fifth European Conference on Machine Learning*, pages 151–163, 1991.
2. Peter Clark and Stan Matwin. Using qualitative models to guide inductive learning. In Paul Utgoff, editor, *Proceedings of the 10th International Conference on Machine Learning*, pages 49–56. Morgan Kaufmann, 1993.
3. William W. Cohen. Fast effective rule induction. In *Proceedings of the Twelfth International Conference on Machine Learning*, 1995.
4. Silvia Coradeschi and Lars Karlsson. A decision-mechanism for reactive and cooperating soccer-playing agents. Electronic Articles in Computer and Information Science 1, Linköping University, 1997.
5. Silvia Coradeschi and Lars Karlsson. A behavior-based approach to reactivity and coordination: a preliminary report. In *Intelligent Agents*, LNAI. Springer-Verlag: Heidelberg, Germany, 1998.
6. Silvia Coradeschi and Lars Karlsson. A role-based decision-mechanism for teams of reactive and coordinating agents. In *RoboCup 97*, LNAI. Springer-Verlag: Heidelberg, Germany, 1998.
7. H. Kitano, M. Tambe, P. Stone, S. Coradeschi, H. Matsubara, M. Veloso, I. Noda, E. Osawa, and M. Asada. The RoboCup synthetic agent challenge. In *Proceedings of the Fifteenth International Joint Conference on Artificial Intelligence (IJCAI-97)*, Yokohama, Japan, August 1997.
8. H. Matsubara, I. Noda, and K. Hiraki. Learning of cooperative actions in multi-agent systems: A case study of pass play in soccer. In *Adaptation, Coevolution and Learning in Multiagent Systems: AAAI Spring Symposium*, pages 63–67, 1996.
9. S.K. Murthy, S. Kasif, and S. Salzberg. A system for induction of oblique decision trees. *Journal of Artificial Intelligence Research*, 2(1):1–32, 1994.
10. Itsuki Noda, Hitoshi Matsubara, and Kazuo Hiraki. Learning cooperative behavior in multi-agent environment: A case study of choice of play-plans in soccer. In *Proceedings of PRICAI'96*, pages 570–579, Cairas, Australia, August 1996.
11. Peter Stone and Manuela Veloso. A layered approach to learning client behaviors in the RoboCup soccer server. *Applied Artificial Intelligence*, 12, 1998.
12. Peter Stone and Manuela Veloso. Towards collaborative and adversarial learning: A case study in robotic soccer. *International Journal of Human Computer Studies*, 48, 1998.

13. Peter Stone and Manuela Veloso. Using decision tree confidence factors for multi-agent control. In Hiroaki Kitano, editor, *RoboCup-97: The First Robot World Cup Soccer Games and Conferences*. Springer-Verlag: Heidelberg, Germany, 1998.
14. Milind Tambe. Towards flexible teamwork. *Journal of Artificial Intelligence Research*, 7:83–124, 1997.
15. Paul E. Utgoff, Neil C. Berkman, and Jeffery A. Clouse. Decision tree induction based on efficient tree restructuring. *Machine Learning*, pages 1–49, 1998.
16. Manuela Veloso, Peter Stone, and Kwun Han. The CMUnited-97 robotic soccer team: Perception and multiagent control. Technical report, Carnegie Mellon Univesity, October 1997.

# Rectangles and Circles: Towards Realistic Simulation of Robots Playing Soccer

Laurent MAGNIN[†]

Electrotechnical Laboratory
1-1-4 Umezono
Tsukuba, Ibaraki
305 Japan
http://ci.etl.go.jp/~magnin
magnin@etl.go.jp

[†]Also affiliated with Lip6 (University Paris VI, France)

**Abstract.** Some simulators for soccer robotics have already been developed. We propose a simulation in which robots are represented by rectangles, not by usual circles. The simulator we use is based on a new model of simulation, called SIEME. This model provides an easy way to describe its entities (e.g. the robots): an entity owns attributes described by mathematical formulas. These formulas can be time dependent. The interactions between the entities are defined by "environmental rules". Our algorithm that applies these rules is based on a discrete-event system simulation. We present some of the equations used in environmental rules to simulate interactions between robots and a ball. A first implementation of SIEME and the soccer simulation has been programmed in Smalltalk and validated by a real-time simulation. We are currently reimplementimg a version of this simulator in Java. It will serve as the basis for the soccer server used in future RoboCup competitions.

## 1. Introduction

The purpose of our research is to create computer simulation of robots playing soccer [11, 17]. In this paper we describe the simulation of the physical interactions between robots, the ball and the playing field. However, the robot's behaviour programming will not be addressed in the present report.

Due to the appearance of the real robots, they must to be represented by rectangles, not by circles as in [14, 16]. This improvement introduces some difficulties: first, we have to deal with more complex geometrical equations. Some of them will be presented in this paper. The second difficulty is that the space discretization associated with "step-by-step" simulation is unsuitable with rectangular robots [12]. For example, it is very difficult to solve the configuration of Fig. 1 in a discretized space:

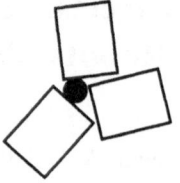

**Fig. 1.** Three Robots Kicking a Ball

We will describe here a solution to these two difficulties. This solution is based on our own "multi-agents" [7] model of simulation (called SIEME) [12, 13][1]. This is the reason why we will first shown this model in the part 2, illustrated by the basis of our robot soccer simulation model. This model gives new kinds of description for the entities (§ 2.1) and the interactions between them (§ 2.2). The simulation algorithm used by SIEME is shortly presented in part 2.3. Subsequently we will present (§ 3) some of the equations we use to modelise the soccer robots and the ball. Finally, two implementations of soccer robot simulations (in Smalltalk and Java) based on the SIEME simulator are shown (§ 4).

## 2. Sieme: An Event Based Simulation Model

### 2.1 Entities & Attributes

First, we describe the "entities" such as the robots, the ball, the field, etc. involved in the system we want to simulate. Unlike other models such as [2, 3, 6, 9] SIEME entity attributes are not only values: they can also be mathematical formulas involving time. Therefore the evolution of the entities depends not only on their actions. For example, when an entity has its own speed, it continues to move without action: given an object initially at location $xT$ in time $Tx$ with a speed vector $sx$, its location $x$ at time $t$ is "$x = xT + sx * (t - Tx)$" (*cf.* Fig. 2).

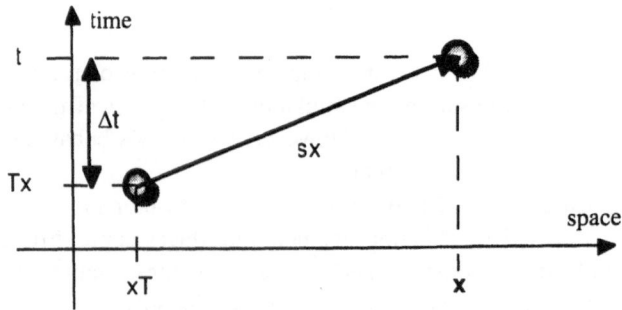

**Fig. 2.** Location depending on time

---

[1] SIEME is a multi-purpose model of simulation, not only dedicated to soccer robots.

The SIEME attribute definition corresponding to this formula is: "x := xT + sx * (t - Tx). "[2] As this example shows, it is possible to use attributes of the same entity ("sx", "xT", etc.) inside an attribute's definition ("x := ..."). Other kinds of attributes are also available: constants and usual variables (i. e. able to be changed during the simulation). "t" is a predefined variable representing the current time. Also due to the attribute "x" declaration, Tx and xT are automatically generated by the simulator.

Shown in the next paragraph is the SIEME'S definition of a ball:

```
R = 5.                    "The ball's radius is a constant"
s.                        "The speed is able to change"
direction.
sx := (direction cos) * s.  "This attribute depends only
   on other attributes"
sy := (direction sin) * s.
x := xT + (sx*(t-Tx)).      "The time value also defines
   the value of the position"
y := yT + (sy*(t-Ty)).
```

We can use almost the same definition for robots, replacing "radius" by "width" and "length":

```
length = 20.
width = 30.
       . . .
```

## 2.2 The environmental rules

The definitions of entities are not enough to describe and simulate our robots playing soccer. We also have to consider the interactions (e.g. collisions) between entities. In a classical simulator, an interaction is only the result of a combination of agents' actions. Our SIEME model goes further ahead and introduces a new concept: the "Environmental Rules" (ER). An ER carries out a computation when a set of conditions becomes true.

Our reference example is "how to simulate collisions between balls." In this case when "two balls are adjacent" becomes true, "change the two balls' direction and speed") is performed by the *ad hoc* ER. This ER recognises the adjacency of circles by using the formula "$\Delta x^2 + \Delta y^2 = (r1 + r2)^2$."

---

[2] The SIEME syntax is based on Smalltalk [8].

Definition of the ER for collision between balls:

```
Collision : | b1 b2 |

"The entities involved, defined by their class"
b1 : Ball.      b2 : Ball.

"Precondition: check that the two balls are not the
same ! "
! b1 ~= b2

"Condition for adjacency of circles"
?(b1,x - b2,x)^2 + (b1,y - b2,y)^2 =
(b1,R + b2,R)^2.

Actions:  "Any kind of "Smalltalk like" code"

| d |     "d is a local variable"

b1,x fix. b1,y fix. "Fix x to update Tx and xT"
b2,x fix. b2,y fix.

d := b1,direction.
b1,direction := b2,direction
b2,direction := d.
```

In conclusion, an ER is defined using this syntax's pattern:

- Entities involved: e.g. "ball : Circle. robot : Rectangle."
- Conditions:
  — Absolutes (e.g., "ball,name ≠ 'foo'")
  — Depending in fine on the time (e.g., "ball,x = robot,x")
- Actions: programming code to execute

## 2.3 The simulation's algorithm

A SIEME simulation is a discrete event simulation [1, 2, 9] in which "Environmental Rules" generate events. The simulator determines—depending on the condition's equations—when an ER will be true. Then it inserts corresponding events for that time in the SIEME's event queue. The general algorithm is presented in Fig. 3:

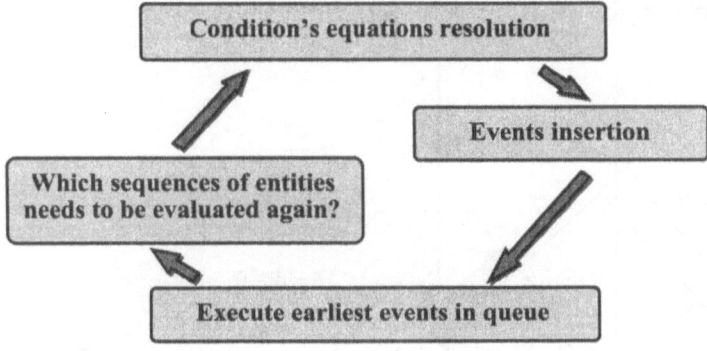

**Fig. 3.** The SIEME algorithm

This algorithm is described in detail in [12, 13].

## 3. Interaction Description

Due to the short nature of this paper we can not give the complete code of environmental rules for our soccer robot simulation. We will concentrate on the main equations used to simulate rectangular robots (using ER).

### 3.1 Robot hiting a side of the field

Fig. 4 to Fig. 6 illustrate the case of a robot touching the right side of the field. The others cases of left, top and bottom sides are described only by analogous formulas (§ 3.1.2). *The formulas shown represent the ER's conditions* (e.g. " $\Delta x^2 + \Delta y^2 = (r1 + r2)^2$ " for the collision between balls).

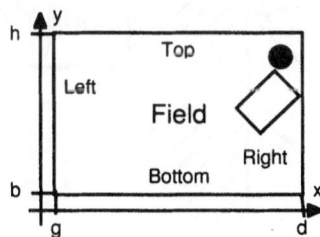

**Fig. 4.** Field's coordinates used inside the equations that will be introduced

## 3.1.1 Field's right side. First configuration: d >= 0

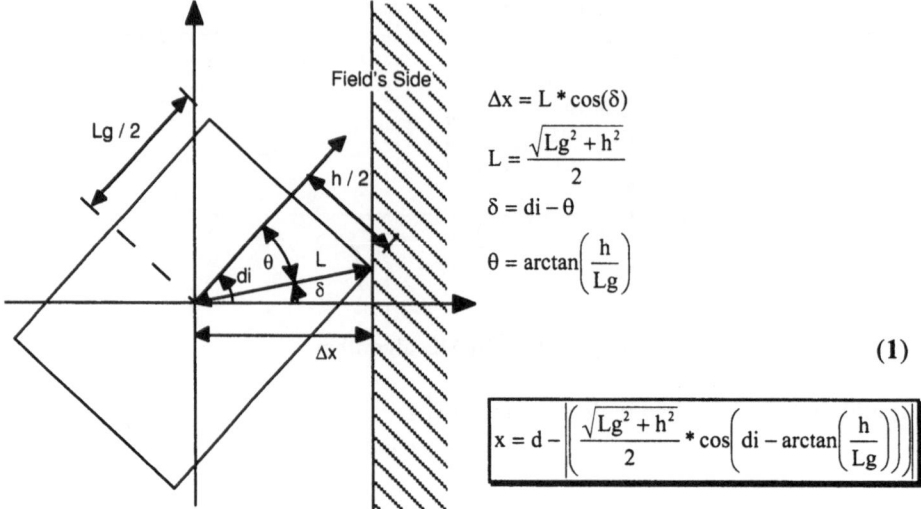

$$\Delta x = L * \cos(\delta)$$

$$L = \frac{\sqrt{Lg^2 + h^2}}{2}$$

$$\delta = di - \theta$$

$$\theta = \arctan\left(\frac{h}{Lg}\right)$$

(1)

$$x = d - \left| \left( \frac{\sqrt{Lg^2 + h^2}}{2} * \cos\left( di - \arctan\left(\frac{h}{Lg}\right) \right) \right) \right|$$

**Fig. 5.** In this case, the equation (1) becomes true when the robot touches the field.

## 3.1.2 Field's right side. Second configuration: d < 0

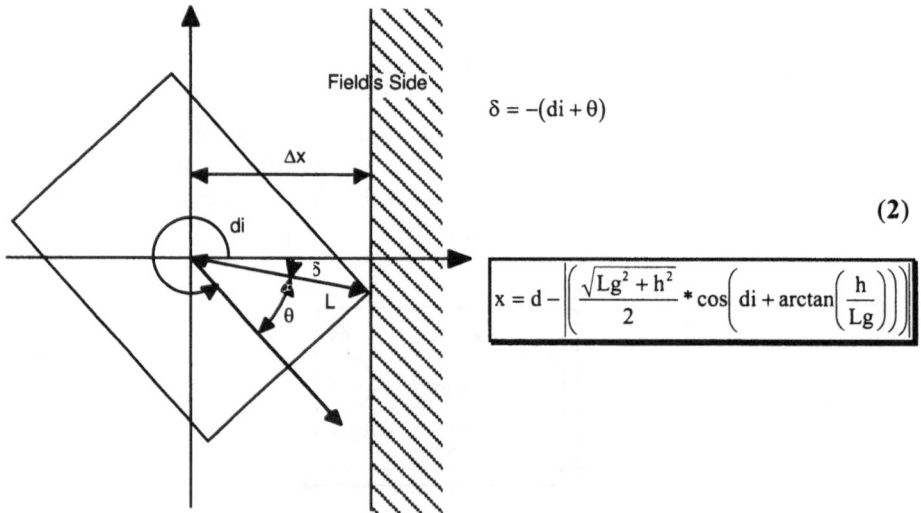

$$\delta = -(di + \theta)$$

(2)

$$x = d - \left| \left( \frac{\sqrt{Lg^2 + h^2}}{2} * \cos\left( di + \arctan\left(\frac{h}{Lg}\right) \right) \right) \right|$$

**Fig. 6.** In this second case it is the equation (2) that becomes true when the robot touches the field.

### 3.1.3 Other cases

For the field's left side:

$$x = g + \left| \left( \frac{\sqrt{Lg^2 + h^2}}{2} * \cos\left( di - \arctan\left( \frac{h}{Lg} \right) \right) \right) \right| \quad (3)$$

or:

$$x = g + \left| \left( \frac{\sqrt{Lg^2 + h^2}}{2} * \cos\left( di + \arctan\left( \frac{h}{Lg} \right) \right) \right) \right| \quad (4)$$

For the field's top side:

$$y = h - \left| \left( \frac{\sqrt{Lg^2 + h^2}}{2} * \sin\left( di \pm \arctan\left( \frac{h}{Lg} \right) \right) \right) \right| \quad (5)$$

For the field's bottom side:

$$y = b + \left| \left( \frac{\sqrt{Lg^2 + h^2}}{2} * \sin\left( di \pm \arctan\left( \frac{h}{Lg} \right) \right) \right) \right| \quad (6)$$

### 3.2 Robot kicking ball

- Case 1: Ball hitting the robot's front or back side:

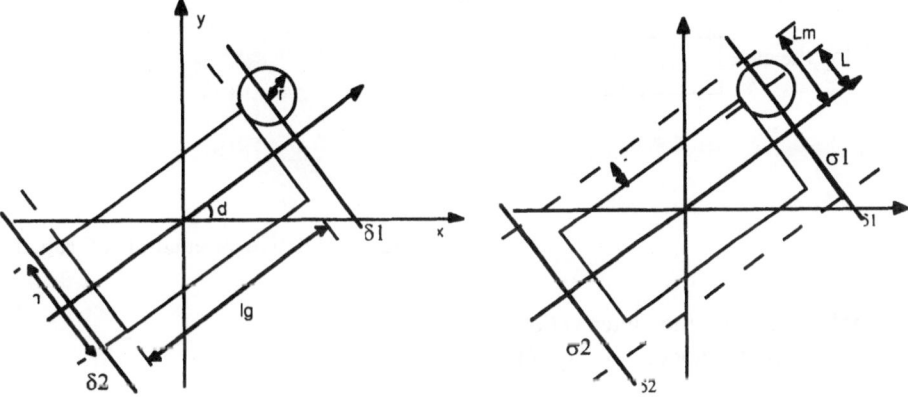

**Fig. 7.** Two geometrical representations used to describe a ball touching a robot's front

$$\left( \left( \left( \cos(d) * \Delta x + \sin(d) * \Delta y = \frac{Lg}{2} + r \right) \vee \left( \cos(d) * \Delta x + \sin(d) * \Delta y = -\left( \frac{h}{2} + r \right) \right) \right) \right. \quad (7)$$
$$\left. \wedge \left( |\cos(d) * \Delta y - \sin(d) * \Delta x| \leq \frac{h}{2} + r + \varepsilon \right) \right)$$

We use an $\varepsilon$ to be sure that this formula is true when the ball hits a robot's corner.

- Case 2: Ball hitting a robot's side:

$$\left[\left(\left(\sin(d)*\Delta x - \cos(d)*\Delta y = \frac{h}{2} + r\right) \vee \left(\sin(d)*\Delta x - \cos(d)*\Delta y = -\left(\frac{h}{2}+r\right)\right)\right) \\ \wedge \left(|\cos(d)*\Delta y + \sin(d)*\Delta x| \le \frac{Lg}{2} + r + \varepsilon\right)\right] \tag{8}$$

## 3.3 Collision between two robots

We check the intersection of two rectangles (i.e. when two robots touch) by checking the following four intersections of one rectangle and one stripe:

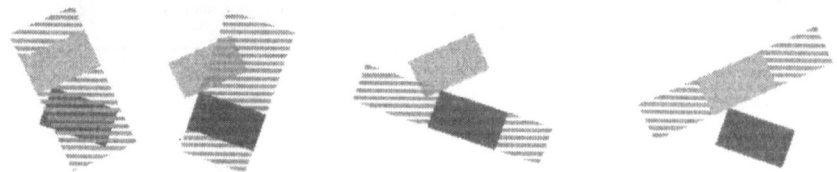

There are two cases depending on φ, which is the relative orientation of robots (cf. Fig. 8):

We develop the case $\boxed{\dfrac{\pi}{2} \le \varphi \le \dfrac{3\pi}{2}}$.

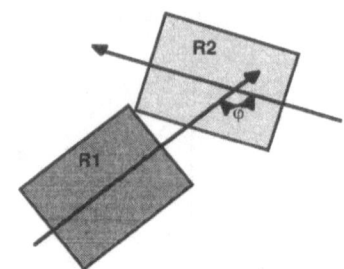

**Fig. 8.** Relative orientation of robots

Also, only one of the intersection's equations will be developed (cf. Fig. 9):

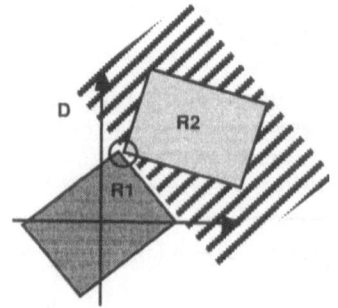

**Fig. 9.** One of the four kinds of intersections

131

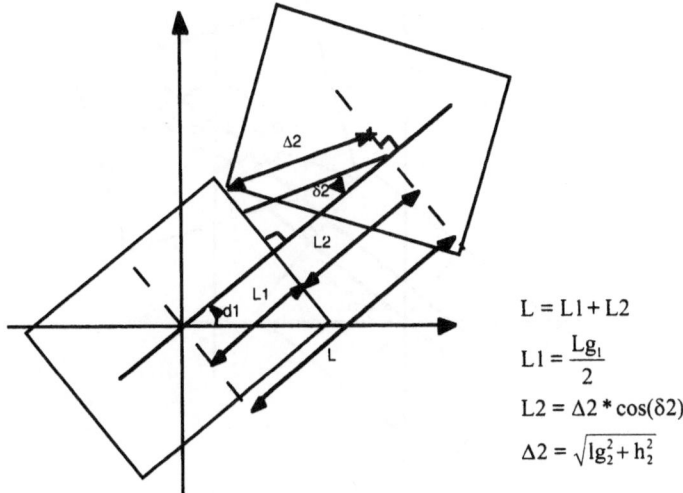

$$L = L1 + L2$$
$$L1 = \frac{Lg_1}{2}$$
$$L2 = \Delta 2 * \cos(\delta 2)$$
$$\Delta 2 = \sqrt{lg_2^2 + h_2^2}$$

**Fig. 10.** First geometrical figure dedicated to describe two rectangles touching themselves using mathematical formulas as in Fig. 9. It represents the distance L between the two rectangles' centres.

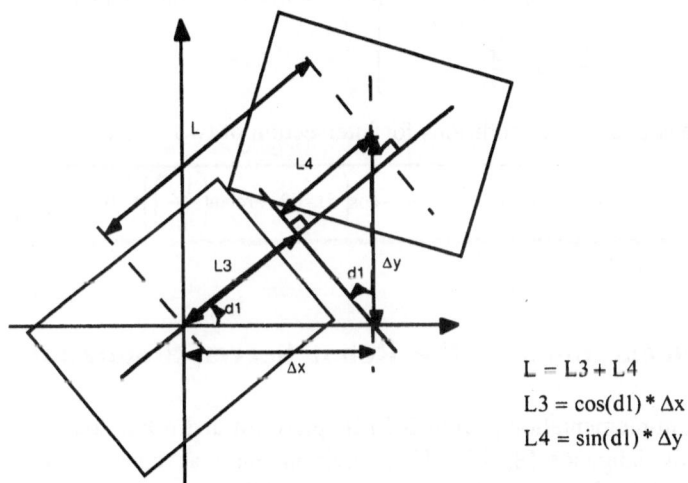

$$L = L3 + L4$$
$$L3 = \cos(d1) * \Delta x$$
$$L4 = \sin(d1) * \Delta y$$

**Fig. 11.** Another representation of Fig. 9's case. It uses another formula to describe the distance L between the two rectangles' centres.

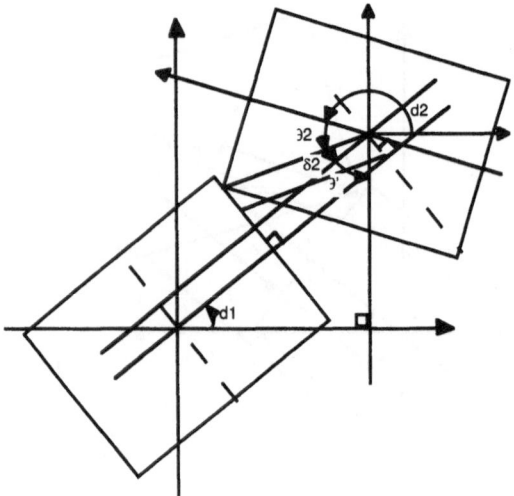

**Fig. 12.** This last geometrical representation of the Fig. 9's case helps to compute the δ2 value used in Fig. 11.

$$d2 + \theta2 + \delta2 + \theta' = \frac{3\pi}{2} \left. \begin{array}{l} \\ \theta2 = \arctan\left(\dfrac{h_2}{\lg_2}\right) \\ \\ d1 + \theta' = \dfrac{\pi}{2} \end{array} \right\} \Rightarrow \delta2 = d1 - d2 - \arctan\left(\frac{h_2}{\lg_2}\right) + \pi$$

Finely, here is one of the conditions for intersection of two rectangles:

$$2 * (\cos(d1) * \Delta x + \sin(d1) * \Delta y) = \lg_1 - \cos\left(d1 - d2 - \arctan\left(\frac{h_2}{\lg_2}\right)\right) * \sqrt{\lg_2^2 + h_2^2} \tag{9}$$

# 4. Implementations of the Robot Soccer Simulation

Our current implementation [12] of SIEME platform and robot soccer simulator uses the Smalltalk language [8, 15]. This environment was chosen for the following reasons: portability[3], fast development, meta-compiler.

This implementation was used in the MICROB research project's context [4]. Despite Smalltalk's slow floating-point computations, this platform was sufficiently efficient to perform real-time simulations[4] (due to the intrinsic efficiency of SIEME algorithm: no need to evaluate each step the interactions between all of the entities).

---

[3] Portability between Macintosh, UNIX station and PC.
[4] In the case of Microb: real-time with 8 robots on a PowerMacintosh.

We present here as example of an MICROB simulation, two robots kicking a ball (cf. Fig. 13):

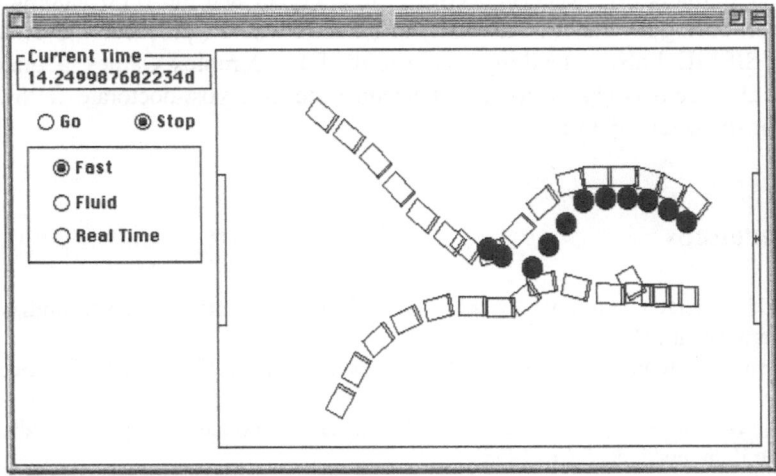

**Fig. 13.** MICROB simulation: pass between robots coming from the left. The numerical value shown in the "Current Time" box is the precise date of the last performed event.

To offer Web interface and a faster portable simulation platform, we are currently developing a new SIEME simulator in Java [10]. This implementation will form the basis of a new version of the soccer server [14] used in the RoboCup [OSA 96, 11] competitions. Nevertheless our new goal is to simulate real soccer players, not robots. This is the reason why this new version represents players by circles instead of rectangles.

## 5. Conclusion

The sharp physical interactions between robots modeled by rectangles in our soccer robot simulation provides new challenges for the larger field of multi-agent systems. Therefore in this paper we propose a new simulation model—using events—based on continuous time and space. One of the improvements of our new model is the ability to use formal descriptions of the interactions between entities (e.g. the ball, the robots, etc.). Some of the equations for the simulation of soccer robotics are presented here. They have been validated by an implementation using Smalltalk language. The next generation of the soccer server used in RoboCup competitions will use this model of simulation in Java language.

# 6. Acknowledgments

I would like to thank my thesis supervisor Prof. J. Ferber and the Lip6 laboratory for their support during my stay in University of Paris VI where I produced the Smalltalk version of SIEME. I also extend my thanks to the ETL Complex Games laboratory— especially Dr. Itsuki Noda—who have welcomed me as a post-doctorate fellow, and JISTEC for financial support.

# 7. References

1. Scott D. Anderson, Multiple Event Stream Simulator, "http://earhart.cs.umass.edu/research/mess.html".
2. Jerry Banks & John S. Carson II, Discrete-Event System Simulation, Prentice-Hall, 1984.
3. Renaud Cazoulat, Modélisation et simulation de la dynamique de population d'agents, Thèse de l'université de Caen, 1995
4. Anne Collinot, Alexis Drogoul & Philippe Benhamou, "Agent Oriented Design of a Soccer Robot Team", Second International Conference on Multi-Agent Systems (ICMAS 96) , pp. 41-47, 1996.
5. Claude Delaye & Laurent Magnin, Systèmes multi-agents, simulations et règles environ-nementales (L'expérience Plages), Journée "Systèmes multi-agents", Paris, P.R.C.-G.D.R. Intelligence Artificielle, 1994.
6. E. Durfee & T. Montgomery, MICE: A Flexible Testbed for Intelligent Coordination Experiments, Ninth Workshop on Distributed Artificial Intelligence, Rosario Resort, Eastsound, Washington, pp. 25-40, 1989.
7. J. Ferber, Les systèmes multi-agents, vers une intelligence collective , InterEditions, 1995.
8. Smalltalk-80 : the language, editors Adele Goldberg & David Robson, Addison-Wesley, 1989.
9. Z. Guessoum & M. Dojat, A Real-Time Agent Model in an Asynchronous Object-Oriented Environment, MAAMAW'96.
10. Java Computing, " http://www.sun.com/java/".
11. Kitano & al., "RoboCup. A Challenge problem for AI", AI Magazine, Vol. 18, no 1, pp 73-85, Spring 1997.
12. Laurent Magnin, Modélisation et simulation de l'environnement dans les systèmes multi-agents : application aux robots footballeurs, Thèse de l'université Paris VI, 1996.
13. Laurent Magnin, SIEME: an Interactions Based Simulation's Model. ESM '98 confer-ence, Manchester, 1998.
14. Itsuki Noda, Soccer Server System, "http://ci.etl.go.jp/~noda/research/kyocho/soccer/server.html".
15. ParcPlace-Digitalk Homepage, "http://www.parcplace.com".
16. Tucker Balch, JavaSoccer, "http://www.cc.gatech.edu/grads/b/Tucker.Balch/JavaBots/JavaSoccer".
17. Osawa & al., "RoboCup: The Robot World Cup Initiative", Second International Con-ference on Multi-Agent Systems (ICMAS 96) , 1996.

# Collective Search by Mobile Robots Using Alpha-Beta Coordination

*Steven Y. Goldsmith and Rush Robinett,III*
*Sandia National Laboratories[1]*
*Albuquerque, NM 87185-5800, U.S.A.*
*Phone: 505-845-8926*
*sygolds@sandia.gov, rdrobin@isrc.sandia.gov*

**Keywords:** collective robotics, collective search, reactive coordination, emergent behavior, collaborative agents

**Abstract.** One important application of mobile robots is searching a geographical region to locate the origin of a specific sensible phenomenon. Mapping mine fields, extraterrestrial and undersea exploration, the location of chemical and biological weapons, and the location of explosive devices are just a few potential applications. Teams of "robotic bloodhounds" have a simple common goal; to converge on the location of the source phenomenon, confirm its intensity, and to remain aggregated around it until directed to take some other action. In cases where human intervention through teleoperation is not possible, the robot team must be deployed in a territory without supervision, requiring an autonomous decentralized coordination strategy. This paper presents the *alpha-beta* coordination strategy, a family of collective search algorithms that are based on dynamic partitioning of the robotic team into two complementary social roles according to a sensor-based status measure. Robots in the *alpha* role are risk-takers, motivated to improve their status by exploring new regions of the search space. Robots in the *beta* role are motivated to improve but are conservative, and tend to remain aggregated and stationary until the alpha robots have identified better regions of the search space. Roles are determined dynamically by each member of the team based on the status of the individual robot relative to the current state of the collective. Partitioning the robot team into alpha and beta roles results in a balance between exploration and exploitation, and can yield collective energy savings and improved resistance to sensor noise and defectors. Alpha robots waste energy exploring new territory, and are more sensitive to the effects of ambient noise and to defectors reporting inflated status. Hypothetically, beta robots conserve energy by moving in a direct path to regions of confirmed high status. Beta robots also resist the effects of noise and defectors by averaging status data, but must rely on alpha robots to improve their performance. Alpha-beta is a reactive strategy that requires directed communication of instantaneous sensor data among team members, but does not rely on a domain model. Alpha-beta coordination is a new and ongoing research effort. We present the basic concepts behind the alpha-beta strategy and exhibit preliminary simulation data that illustrate the collective search modes in an idealized search domain.

[1] Sandia is a multiprogram laboratory operated by Sandia Corporation, a Lockheed Martin Company, for the United States Department of Energy under Contract DE-AC04-94AL85000.

# 1 Introduction and Motivation

Many challenging new applications in robotics involve distributed search and sensing by a robotic team. Mapping mine fields, extraterrestrial and undersea exploration, exploring volcanoes, the location of chemical and biological weapons, and the location of explosive devices are just a few. This paper presents initial but ongoing research into the issues of collective and emergent behaviors in teams of mobile robots tasked with locating specific sensory phenomena. Our motivation for this line of inquiry is the engineering and eventual deployment of large numbers of inexpensive, expendable sensory robots in hazardous or hostile environments, with a particular emphasis on sensing concentrations of hazardous chemicals in terrestrial environments. The problem suite of interest involves the most demanding of sensing environments; rough terrain with obstacles, non-stationary and dilute chemical concentrations, deliberate interference by hostile robots, and limited opportunities for human interaction with the robots through teleoperation [Klarer 1998]. Because human intervention is not always possible in these environments, decentralized coordination schemes which feature collective decision-making by individual autonomous robots are the most promising avenues of research. Overcoming the limitations of crude but inexpensive chemical sensors by using distributed signal processing algorithms that utilize shared data from a large number of agents is another important concept to be investigated. An important issue not generally addressed in robotics research is deliberate and subtle interference with the goals of the robotic team by impostor robots.

Geographical search problems that use robotic teams can be divided into three broad classes: source identification, source mapping, and source localization [Goldsmith and Robinett 1998a]. Robots performing source identification must answer the question "Does region R contain phenomenon X"? A simple yes or no is an adequate answer, and the task can in principle be accomplished by a robot team without actually localizing the target phenomenon. Source mapping requires the robot team to perform an exhaustive search of an area and to localize all phenomena within the region. Source localization problems require precise localization of a target source within a given region. In the simplest form of the source localization problem, a single sensible source is present somewhere within the search space. The search space is divided into two regions based on the quality of sensor data available in the region. The *insensate region* is characterized by a low signal-to-noise ratio. Robots roaming in this region are without information to guide their search activities, and effective search requires multi-agent coordination mechanisms that involve explicit collaboration [Cao, Fukunaga, Kahng, and Meng 1993]. Some coordination strategies for organized collaborative search in zero-information environments are discussed in [Spires & Goldsmith 1998]. The *sensate region* contains the source and is characterized by a signal-to-noise ratio significantly greater than unity. Robots operating in the sensate region have usable but noisy sensory information to guide their search.

Designing a mobile robot team to search a sensate region for a specific target phenomenon involves numerous engineering tradeoffs among robot attributes and environmental variables. For example, battery-driven robots have a finite energy store and can only search a finite area before running down. Success at finding a target source with finite energy resources depends on other characteristics of the robot such as sensor accuracy and noise and efficiency of the locomotive subsystem, as well as properties of the collective such as the number of robots in the team, the use of shared information to reduce redundant search, and the team coordination strategy used to ensure a coherent search process.

## 2 Alpha-Beta Coordination

This paper is concerned with solving the source localization problem using a decentralized coordination strategy we call *alpha-beta coordination*. The *alpha-beta* coordination strategy is a family of collective search algorithms that allow teams of communicating agents[2] to implicitly coordinate their search activities through a division of labor based on self-selected roles and social status. In an alpha-beta team, an agent plays one of two complementary roles. Agents in the *alpha* role are motivated to improve their status by exploring new regions of the search space. Agents in the *beta* role are also motivated to improve, but are conservative and tend to remain aggregated and stationary until the alpha agents have clearly identified better regions of the search space. An agent selects its current role dynamically based on its current status value relative to the current status values of the other team members. Status is determined by some function of the agent's sensor readings, and is generally a measurement of source intensity at the agent's current location. An agent's decision cycle comprises three sequential decision rules: (1) selection of the current role based on the evaluation of the current status data; (2) selection of a specific subset of the current data; and (3) computation of the next heading using the selected data. Variations of these decision rules produce different versions of alpha and beta behaviors that lead to different global properties.

Partitioning the robot team into alpha and beta roles produces a balance between exploration and exploitation. Alpha agents waste energy exploring low-status regions of the search space, but communicate valuable state information to team members that prevents costly reexploration of low-status regions. Alpha agents by nature seek to emulate and ultimately surpass the highest-performing team members and are therefore more sensitive to the effects of transient noise and are more susceptible to the influence defectors[3] reporting false status values. Beta agents use energy wisely by resisting transient influences and moving in a direct path to high-status regions of the search space identified by alpha agents. Hypothetically, beta agents resist noise and defectors (we do not provide support for this claim herein) by selective re-sampling and averaging of status data, but must rely on alpha robots to improve their performance. Consequently, beta agents can be mislead by noise and defectors under some circumstances through second order effects if many of the the alpha agents are mislead. Alpha-beta coordination relies on the following assumptions:

---

[2] We will use the term agent hereafter to signify the generality of the alpha-beta concept and to stress that we have not yet implemented the technique on actual robotic vehicles.

[3] Defectors may inadvertently misrepresent their status because of flaws, or may be impostors that deliberately attempt to mislead the loyal team members. These kind of effects can be characterized as Byzantine failures [Lamport, Shastak, and Pease 1982].

1. Team members have a reliable communications mechanism.

2. The team is positioned in the (noisy) sensate region surrounding a target source.

3. The terminal goal of the team is to converge on the source target.

4. A higher status value implies a higher probability that the source is located near the corresponding coordinates.

Alpha-beta agents are *eusocial* [Mcfarland 1994] in nature; agents must cooperate to succeed. Agents always broadcast their most current sensor data as a normative behavior. An agent's model of the environment is based solely on their current local sensory data and the current shared data obtained from the other members of the team. Individual agents have no sensor memory and consequently cannot locate a source alone. As such, the alpha-beta strategy is a reactive collective search strategy rather than a collaborative strategy. Agents are implicitly cooperative, and do not use explicit forms of collaboration. The alpha-beta strategy is a behavior-based control strategy closely related to the approach of Mataric [1994]. Alpha-beta teams behave in a manner similar to that of of simple insect societies [Kube and Zhang 1993]. Alpha-beta agents search without centralized leadership or hierarchical coordination. The primary collective mode of an alpha-beta team is to aggregate in a region of high-intensity, without any other objectives. Alpha-beta teams are robust to single-point fail-stop failures in team members; agents simply use the latest data transmitted by other team members without regard to the identity of the sender. Alpha-beta coordination requires a minimum of knowledge about the search environment. Agents have no prior assumptions about the nature of the intensity surface, its spatial coherence, gradient field, or any other analytical information. As such, the alpha-beta strategy is intended to be as general-purpose and as assumption-free as possible. In formulating the alpha-beta strategy, we have carefully constructed the problem context and agent capabilities to focus the research in a particular direction, namely away from traditional symbolic AI approaches and towards the dynamical systems and behavior-based/emergent behavior approach. This paper presents the alpha-beta concept and exhibits preliminary simulation data in an ideal environment. Our goal is to demonstrate collective coordination based on self-selected dynamical control laws that change in response to the collective state of the team.

## 3 Alpha Beta Coordination Algorithms

A full mathematical treatment of alpha-beta coordination is in progress [Goldsmith & Robinett 1998b] but is beyond the scope of this paper. The current state-space formulation comprises a system of non-linear, time-varying difference equations of order N, where N is the instantaneous number of agents. The issues of primary importance are stability, energy efficiency, convergence, and steady-state localization error.

A simple social metaphor provides an intuitively satisfying if imprecise description of the basis for alpha-beta coordination algorithms. The cohesion of an alpha-beta society is based on a common normative goal: each agent is motivated to improve its social status by associating with other agents of higher status. Social status is determined by a scalar function of the shared sensor data communicated by other agents. The only assumption underlying alpha-beta algorithms is that the status function orders points in the search space according to the probability that a source is located at the point. On each decision cycle, each agent broadcasts it current social status as a scalar value, $s_i$, along with a location vector, $v_i$, to all other agents, and receives their status values in return. An agent attempts to improve its standing through emulation by moving to a region occupied by agents reporting superior status. This simple goal pressures agents to: (1) aggregate into groups; and (2) to aggregate in the region of highest known status. To determine its next destination, each agent first computes the common ordered set $V=\{v_i\}$ according to the linear ordering ($\leq$) of agents provided by the status readings $S=\{s_i\}$[4]. The agent uses $S$ to partition its fellow agents into two castes. The *alpha caste* is the set $A_0$ of all agent positions corresponding to agents that have a social standing superior to agent $a_0$: $A_0 = \{v_k|s_k > s_0\}$. The *beta caste* $B_0$ is the set of all agent positions corresponding to agents with lower social standing than agent $a_0$: $B_0 = \{v_k|s_k > s_0\}$. The beta set $B_0$ includes agents of equal status because an agent always seeks to improve its current status. There are a variety of approaches to using the alpha and beta sets to generate the agent's next heading. The vectors in the set $A_0$ can be used to influence the agent to move towards its members, creating a social pressure to improve called *alpha-pull*. The vectors in the set $B_0$ can be used to influence the agent to move away from its members, creating a second social pressure to improve called *beta-push*. Either set or $V$ itself can be used in a variety of ways to provide pressure to aggregate. Alpha-pull and beta push are heuristic in nature and do not necessarily lead to average improvement in arbitrary environments. Designing and testing different decision rules based on the data vectors in $V$, $A_0$, and $B_0$, or subsets thereof, is the means for investigating the different global behaviors of alpha-beta teams.

A special case of importance is when $V=A_i=B_i$. In this case every agent has identical status, corresponding to the zero-information (maximum information entropy) state previously mentioned. When a zero-information state is detected, the team can disperse to broaden the search area by using beta-push (all members are in the beta sets of all other members) to compute a trajectory that leads the agents on the outer edges of the cohort region away from the team's centroid. As the density of the team decreases, more agents are free to move away from the centroid, eventually resulting in a dispersed team. A minimum limit on team density prevents the ultimate loss of team coherence. If the team members cannot find the sensate region, they must resort

---

[4] The unordered set of of readings can be used to compute the obvious non-uniform gradient estimates. We have investigated gradient search algorithms and use them as a baseline for comparison of alpha-beta performance. Some forms of alpha-beta algorithms currently under investigation use gradient estimates for alpha decisions.

to a collaborative search mode as mentioned previously.

If $V=\emptyset$, the agent is alone. For the purposes of this research, agents that lose contact with the team remain immobilized. This "hug a tree" philosophy saves energy but may not lead to a reunion with the team and to eventual arrival at the target source. A variety of possible solo behaviors will be investigated later, including random search, gradient search, and using the last known heading to determine the agent's trajectory.

The general form of the alpha-beta update rule uses a linear combination of the vector data in $V$:

$$v_i(k+1) = v_i(k) + a(k)[v(k) - v_i(k)] \tag{1}$$

where $a$ is a weighting vector derived from the application of some scalar function to the current status measurements $S$ corresponding to the vectors in $v$. The nature of the function applied to $S$ and the specific subset of vectors selected from $V$ determine the group behavior exhibited by this version of alpha-beta teaming.

The alpha set $A$ contains a distinguished subset of elements: the agent or agents with the highest status value. An agent with the highest status in the cohort has no alpha caste; $A=\emptyset$. These agents are the *ø-alpha* agents and cannot experience alpha-pull. The choice of a decision rule for a *ø-alpha* agent is limited two possibilities: (1) don't move; and (2) move away from the team along a vector derived from the B-vectors (beta-push). In the first option, the ø-alpha[5] identifies the location of highest known status and acts as a stationary beacon for the rest of the team. This is a conservative strategy that  saves energy and ensures that the agent remains at the top of the heap, but does not immediately explore the region around the highest intensity reading. The second option uses some form of beta-push to move the ø-alpha away from the team. This is a risky strategy because the status of the ø-alpha may decrease, but it provides more information to the team and can possibly shorten convergence time.

The beta set $B_0$ also contains a distinguished subset of elements: the agent or agents with the lowest status value. These *ø-beta* agents represent the social floor of the team, and always use some form of alpha-pull to improve their status.

The remaining members of the cohort have non-empty alpha and beta sets. Such an agent can experience the effects of both alpha-pull and beta-push. There are many possible decision rules for determining the next heading based on the partition $\{A_0, B_0\}$. In general, an agent must decide whether to be radical or conservative in its attempt to improve its status. The approach taken here is to provide three classes of behavior. For an agent team with N agents the update rules are:

1. The ø-alpha agents use the conservative decision mode and remain immobile:
   $v_i(k+1) = v_i(k)$.

2. The m agents in $V$ with the highest status values self-select alpha behavior and use

---
[5] Although there may be more than one ø-alpha , we use the singular hereafter.

the following update rule: $v_i(k+1) = v_i(k) + u[v^*(k) - v_i(k)]$, where $v^*(k)$ is the location of a ø-alpha agent, selected at random, and $\underline{u}$ is a factor that provides pressure to move beyond the alpha agent along a line passing through the points $v^*(k)$ and $v_i(k)$. Note that $u > 1.0$ must hold for improvement.

3. The remaining N-m agents in **V** self-select beta behavior and use the following update rule: $v_i(k+1) = v_i(k) + a(k)[v(k) - v_i(k)]$, where $v(k)$ are all members of $A_i$, and $a(k)$ is the corresponding vector with elements $a_j = s_j/D$, and

$$D = \sum s_k, \ k=1, N \qquad (2)$$

Under this regime, self-selected alpha agents attempt to exceed the performance of the stationary ø-alpha agent by attempting to overrun it. Self-selected beta agents compute a weighted average of the alpha vectors based on normalized status values and move towards the resultant. A conservative beta agent seeks to improve its status to the average status of its corresponding alpha set by moving to the point of the center-of-mass of the alpha set. This "safety in numbers" approach provides a tendency to aggregate in the most current region of highest known performance, but averages many alpha status positions to reduce noise and the influence of outliers. This behavior provides the beta population with some inertia, but still retains the tendency to improve the status of the population on average.

The important parameters in this regime are u, the "overrun factor" that determines the amount by which an alpha will attempt to move beyond a ø-alpha agent, and the alpha ratio, defined as =m/N, that determines the proportion of alpha agents exploring the search space.

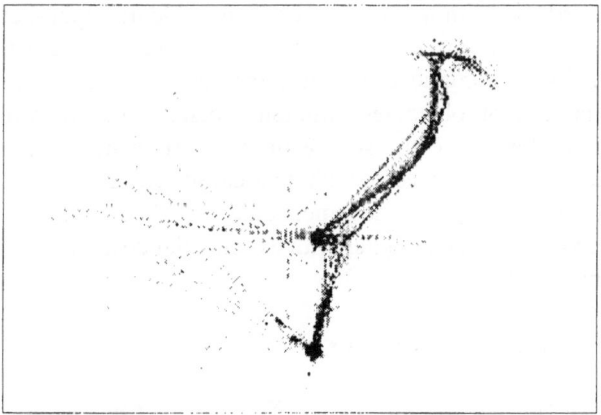

Figure 1: A team of 50 agents start in the upper right and locate a source at the center of the figure. The source intensity drops to zero and agents disperse to the right to locate another source. The source reappears in the lower center and agents once again converge upon it.

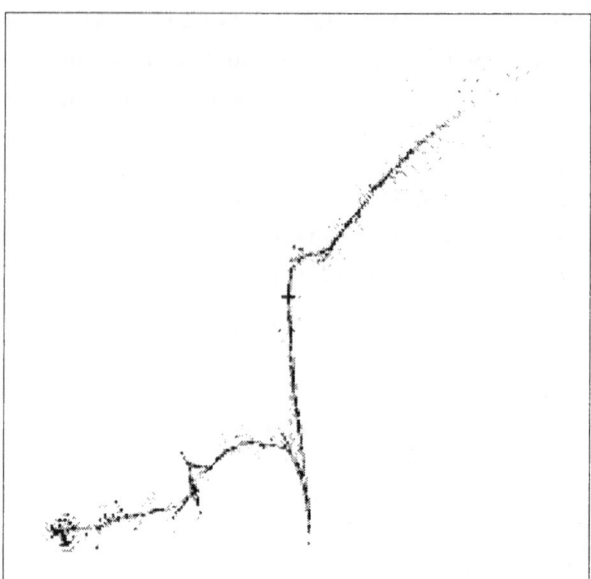

Figure 2: A team of 50 agents start in the upper right and eventually locate a source at the lower left. The annular region around the source results from alpha agents continuously searching around the source. The search trajectory is typical of an alpha-beta agent team.

## 4  Simulations and Results

The alpha-beta coordination strategy was simulated in an ideal 2-D world using ideal agents. The intent of these initial simulations was to study the convergence and coordination properties of alpha-beta rather than evaluate alpha-beta in a realistic environment. The simulations provide a best-case baseline against which various complications such as communications noise and sensor noise can be evaluated later on. The world is free of obstacles, ambient noise, communications errors and convection currents that make the source intensity field non-stationary and time-varying. Ideal agents are point-masses with no area, so crowding is not an issue. Ideal agents have noise-free sensors, and movement on each step is bounded.

The target source was a radial emitter with exponential decay factor $\underline{b}$ and a uniform random noise component $\underline{w}$:

$$Z(r) = w + \exp -(r \bullet b) \tag{3}$$

where r is the radial distance from the origin. The metric of interest for this study is the mean-squared distance from the target, a measure of the team's learning rate and steady-state convergence error. For each simulation run, alpha-beta agents are initially positioned with the same distribution in the x-y plane. A control group comprising agents with identical starting points but with knowledge of the source location provide **a baseline for**

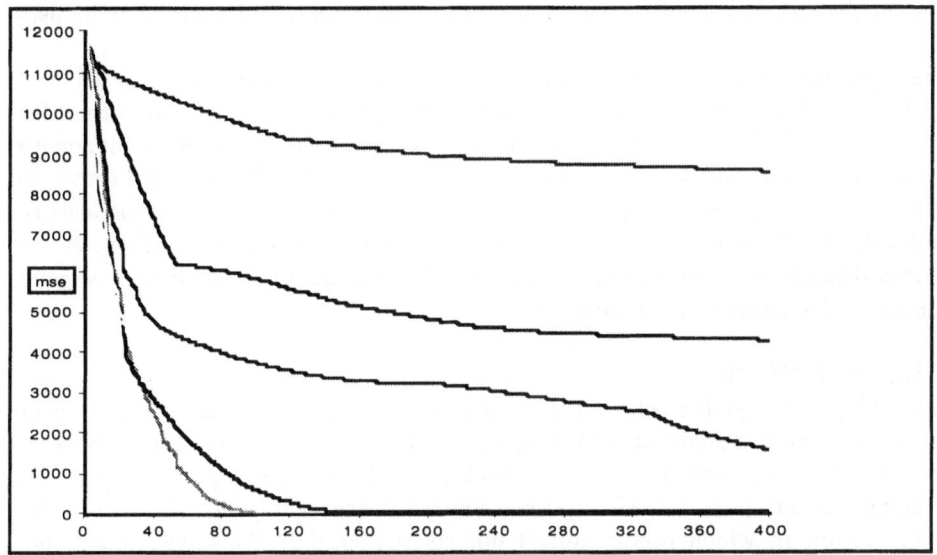

Figure 3. Mean-squared error vs. step for u=2.0 and (1) R=0.1 (upper); (2) R=0.2; (3) R=0.4; (4)R=0.5(lower); (6) R=0.6;(7)R=0.8; (8)R=1.0(third from top).  Convergence is for ß=0.5. Notice the diminishing returns for ß>0.5.

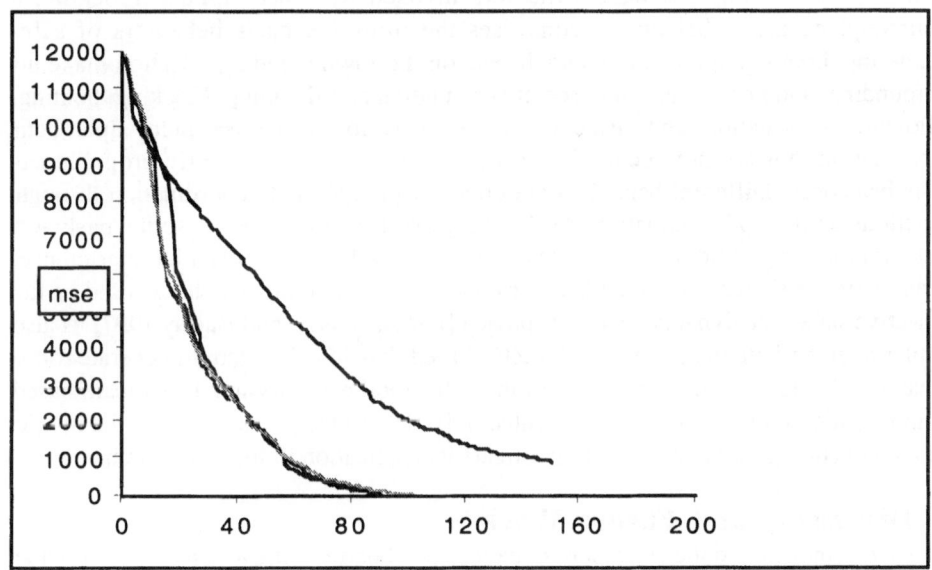

Figure 4: Mean-squared error vs. step for R=0.5 and: (1) u=2.0; (2) u=3.0 ; (3) u=4.0;(4) u=8.0;(5) u=10.0(upper).

learning curve for the team.  Figures 1 and 2 show typical traces of alpha-beta agents.

The simulation results confirm that the team can find a source under ideal conditions.

The alpha ratio is R critical to effective search. A critical mass of alpha agents is needed to influence the beta agents to follow the alpha trajectory. A ratio of not less than 0.4 is needed for reliable search given a u value of 2.0. Maximum convergence rate and minimum steady-state mean-squared error occur at R=0.5. Figure 3 shows the learning curves for various alpha-beta ratios. Convergence rate is somewhat sensitive to the alpha u parameter as expected, favoring greater values of u at the expense of increased steady-state mean squared error. Very large values of u slow the convergence rate and lead to larger steady-state errors.

## 5 Related Work

The emergence of global behavior from local interactions among autonomous agents has been studied extensively. Investigations of collective behavior in robots are considerably more rarefied, and studies involving collective search are rarer still. The foraging problem [Arkin & Hobbs 1993, Goss & Deneubourg 1992, Mataric 1994, Steels 1990], in which robots collect objects scattered in the environment, is a canonical problem related to the source location problem.

The alpha-beta strategy falls squarely in the behavior-based control camp [Brooks 1991, Brooks 1986, Mataric 1992]. Mataric (1994;1995) describes group behaviors in terms of combinations of basis behaviors invoked by sensor inputs. Flocking, a commonplace group behavior, comprises the primitive basis behaviors of safe-wandering, homing, aggregation, and dispersion. Following and aggregation make up surrounding, and herding is composed of surrounding and flocking. Flocking, homing, following, aggregation, and dispersion are all behaviors that arise under alpha-beta coordination, but are not accomplished by compositions of explicitly programmed basis behaviors. Different behaviors are obtained in alpha-beta coordination through variations on the update equation (1).Goldberg and Mataric [1997] describe pack and caste criteria for partitioning a robot team to achieve arbitration of spatial interference. Their approach shares with alpha-beta the concept of behavioral switching based on the collective state. The dynamics-based approach [Large, Henrik, and Bajcsy 1997] is also similar to alpha-beta in its use of of a vector-based dynamical system to generate robot behaviors. Social entropy, a measure of the behavioral diversity in a robot team based on information entropy, has been presented in [Balch 1997]. This is a potential metric for alpha-beta regimes and we will investigate its application in future research.

## 6  Discussion and Future Work

We have demonstrated the concept of dynamic social partitioning as a means to provide collective benefits to an agent team searching for source targets. Initial simulations confirm the ability of the team to find a source and stabilize the steady-state mean-squared error.

Our future research will focus on further investigations of alternative forms of alpha-beta algorithms inspired by molecular dynamics and statistical mechanics. We intend to investigate new forms of interaction rules that are based on non-linear functions of the entire measurement set rather than on partitions of the measurements. We will also investigate dynamic adjustment of the alpha u parameter and the alpha/beta ratio through reinforcement learning techniques under the alpha-beta regime presented in this paper. Simulations involving more realistic environments containing obstacles, convection effects on chemical plumes, and more detailed models of robotic vehicles will be conducted on parallel processors for large numbers of agents if required. Ultimately, we will attempt to implement alpha-beta strategy on actual robotic vehicles.

## Acknowledgements

We wish to thank our colleagues at the Advanced Information Systems Laboratory, Laurence Phillips and Shannon Spires, for technical discussions and comments on this paper, and for coding and debugging portions of the simulation.

## References[6]

Arkin, R. and Hobbs, J. 1994. Dimensions of communication and social organization in multi-agent robotic systems. In *Proc. Simulation of Adaptive Behavior.*

Balch, T. 1997. Social entropy: A new metric for learning multi-robot teams. In *Proc. FLAIRS-97.*

Brooks, R. 1986. A robust layered control system for a mobile robot. *IEEE Journal of Robotics and Automation* RA-2.

Brooks, R. 1991. Intelligence without reason. In *Proc. IJCAI-91.*

Cao, U., Fukunaga, A., Kahng, A., and Meng, F. 1995. Cooperative mobile robotics: Antecedents and directions. In *Proc. of IEEE/RSJ IROS.*

Goldberg, D. and Mataric, M. 1997. Interference as a tool for designing and evaluating multi-robot controllers. In *Proceedings, AAAI-97, Providence, Rhode Island, July 27-31, 1997, 637-642.*

Goldsmith, S. and Robinett, R. 1998a. Collaborative search by mobile robots- Part I: Problem definition. In progress.Preprint at www.sandia.gov/aisl/robotics/collaboration

Goldsmith, S., and Robinett, R. 1998b. Stability and performance of alpha-beta coordination. Working paper. Preprint: www.sandia.gov/aisl/robotics/ab-stability

Goss, S., and Deneubourg, J. 1992. Harvesting by a group of robots. In *Proc. European Conf. on Artificial Life.*

---

[6] The AISL website will be opened to the Internet before the July 4 conference opening.

Kube, C. and Zhang, H. 1993. Collective robotics: From social insects to robots. *Adaptive Behavior, 2(2)*.

Klarer, P. 1998. Flocking small smart machines: An experiment in cooperative multi-machine control. Sandia National Laboratories Report. In progress.

Lamport, L., Shostak, R., and Pease, M. ``The Byzantine Generals Problem,'' *ACM Transactions on Programming Languages and Systems 4, 3* (July 1982), 382--401.

Large, E., Henrik, C., Bajcsy, R. 1997. Dynamic Robot Planning: cooperation through competition. In *Proceedings of the IEEE International Conference on Robotics and Automation*, Albuquerque, NM.

Mataric, M. 1992. Behavior-based systems: Key properties and implications. In *IEEE International Conference on Robotics and Automation, Workshop on Architectures for Intelligent Control Systems*.

Mataric, M. 1994. Interaction and intelligent behavior. *MIT AI Lab Tech report AITR-1495*.

Mataric, M. 1995. Designing and understanding adaptive group behavior. *Adaptive Behavior (4:1)* .

McFarland, D. 1994. Towards robot cooperation. In *Proc. Simulation of Adaptive Behavior*.

Spires, S. , and Goldsmith, S. 1998. Exhaustive Geographic Search with Mobile Robots Along Space-Filling Curves. Submitted to Collective Robotics Workshop, Agent World '98.

Steels, L. 1990. Cooperation between distributed agents through self-organization. In *European Workshop on Modeling Autonomous Agents in a Multi-Agent World*.

# A Knowledge-Level Approach for Building Human-Machine Cooperative Environment

H. Takeda, N. Kobayashi, Y. Matsubara, and T. Nishida

Graduate School of Information Science,
Nara Institute of Science and Technology
8916-5, Takayama, Ikoma, Nara 630-01, Japan
takeda@is.aist-nara.ac.jp
http://ai-www.aist-nara.ac.jp/

**Abstract.** In this paper, we propose the *knowledgeable environment* as a framework for integrated systems for human-machine co-existing space. In the knowledgeable environment, integration is achieved as knowledge-level communication and cooperation. We abstract all machines as information agents whose roles and communication are understandable for people. Ontology is here used as explicit representation of the abstraction of the environment which includes agents, objects, and activities of agents and people. Cooperation is also achieved by using ontology. We re-define concept of human-machine interaction in the light of knowledge-level interaction, i.e., interaction with various logical and spatial relation among participants. We realized a prototype of the knowledgeable environment with two mobile robots, rack and door agents, and demonstrated how cooperation among robots, machines and people could be implemented.

**Keyword:** real-world agent, ontology, mediation, cooperation, human-robot interaction

## 1 Introduction

In recent years, various types of computers and computer-controlled machines are introduced into our daily life. It is not distant future when so-called robots be also introduced there. Such machines are expected to improve quality of our life because they can provide performance which we cannot have but want to enhance. But introduction of these machines currently makes us annoying because of variety of their behavior and interface, i.e, each of them has its behavior and interface to user and requires us to understand them. It implies that there need intelligence that should shorten distance between human and machines and integration that should flatten complicity due to variety of machines. In short, we need a framework for integrated intelligent systems for human-machine co-existing space. In this paper, we propose the *knowledgeable environment* in which integration is achieved as knowledge-level communication and cooperation.

It is a new and challenging field for robotics, artificial intelligence, and human interface domains to deal with space for human activity. One reason for it is dynamics in physical space. Distributed and therefore cooperative systems are needed to capture spatially distributed human activities. The other reason is that human activities cannot intrinsically modeled in computers. It implies that human-machine interaction is an indispensable issue that can bridge human and computer activities. We can summarize these problems as the following three research issues;

**1. Modeling of environments including machines and people**: Its meaning has two-holds. One is to model not only machines and environments but also people. We cannot have perfect models of human activities as we mentioned, but partial models are still important to capture human-machine co-existing space. The other is to make models of environments understandable to humans, i.e., models are not only for machines but also humans. It is natural because people are also participants of the environments for which models are provided.

**2. Extension of basis of human-machine interaction**: Various and distributed sensors to detect human activities and presentation methods to people are needed to realize natural human-machine interaction in human-machine co-existing environment. One approach is to extend variety of physical instruments[9]. The other approach is to extend concept of sensoring and presenting. For example, we can call tracking of movement of people[13] as a sensor. Our three distinction of human-machine interaction (described Section 6) is a proposal for this approach.

**3. Cooperative architecture for real-time distributed problems**: People and machines are spatially distributed and synchronized, It means that two types of spatial distribution and synchronization exist, i.e., those for environments (machines are distributed and synchronized) and those for problems (human activities are distributed and synchronized). We need cooperative systems to integrate machines and people in such situation.

## 2 The Knowledgeable environment

Our approach called the *knowledgeable environment* is aiming to build a framework for cooperative systems for human-machine co-existing space. Figure 1 shows an image of space which we want to realize. In the space, people and machine are mutually cooperative, i.e., people can ask some tasks to machines and vice versa. It may seem strange that machines can ask something to people. It may be possible to assume that machines are almighty in environments like factories because lack of existence of human enables environments to be designed solely for machines. Since environments like our living space cannot be designed solely for machines, some tasks cannot be achieved only by machines but by combination of machines and people. In such case, people can be asked by machines.

In the knowledgeable environment, the above three problems is solved by knowledge-level modeling and interaction. We abstract all machines as *agents* whose roles and communication are understandable for people. Ontology is here used as explicit representation of the abstraction. Cooperation is achieved by

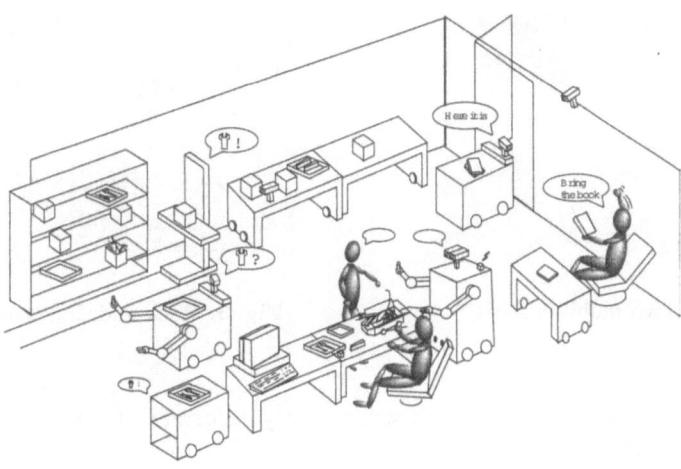

**Fig. 1.** An image for the knowledgeable environment

using ontology. We re-define concept of human-machine interaction in the light of knowledge-level interaction, i.e., interaction with various logical and spatial relation among participants. In this paper, we discuss the following four methods that we are currently developing in order to realize the knowledgeable environment.

1. Agentification of robots and machines
2. Ontology as modeling of environment
3. Ontology-based task mediation for cooperation
4. Various interaction between human and robots from intimate to cooperative

## 3  Agentification of robots and machines

The basic idea of our approach for modeling machines is to model them as software agents that can communicate to each other with some abstracted language. The merit of this approach is as follows;

- Abstracted definition of agents is applicable
- Techniques developed for software agents like cooperation are available
- Cooperation between software agents and machines is solved in a single architecture

We call agents that have facilities to obtain information from the physical environment or to do something to the environment as *real-world agents*. On

**Fig. 2.** Two mobile robots          **Fig. 3.** Rack and door agents

the other hand, we call agents concerning only information in computers as *information agents*.

All robots and machines are agentified as KQML agents[4]. KQML(Knowledge Query and Manipulation Language) is a protocol for exchanging information and knowledge among agents. KQML is mainly designed for knowledge sharing through agent communication. A KQML message consists of a message type called *performative* like *ask, tell* and *subscribe*, and a number of parameters like *sender, receiver, content* and *language*. For example, a message content is written as a value of parameter *content*. We mostly use KIF (Knowledge Interchange Format)[5] as language to describe message contents. KIF is a language for interchanging of knowledge and based on the first-order predicate logic.

A real-world agent can consist of some sub-agents each of which performs specific information processing in the agent. By separating facilities in an real-world agent, we can design agents without depending computational performance of each robot or machine. A typical real-world agent consists of three sub-agents, namely *KQML handling sub-agent* that parses and generates KQML messages, *database sub-agent* that holds status of the agent itself and its environment, and *hardware controlling sub-agent* that sends commands to actuators and obtains sensor values.

We currently agentified a mobile robot with two manipulators called *Kappa1a*, a mobile robot without manipulators called *Kappa1b* (see Figure 2), computer-controlled rack and door(see Figure 3). A manipulator has six degrees of freedom and a gripper.

We also treat human as agents to some extents. We can provide action knowledge that includes human as participants of actions. As we will describe in the next section, ontology is provided as common vocabulary between human and computer agents as well as among agents. But there are many difficulty to treat human just as computer agents. In the context of our definition of agents, difference of way of communication is crucial. Human has various communication channels that are to choose depending on situations, while computer agents have a single channel. It implies that modeling of human-machine interaction as inter-agent interaction is needed. We will discuss how human-machine interaction is modeled and integrated in the knowledgeable environment in Section 6.

# 4 Ontology as modeling of environment

Our aim is to establish an information infrastructure to cooperate heterogeneous real-world agents at knowledge level, i.e., to clarify what knowledge is needed for those agents for cooperation. We introduce ontologies for object, space, and action as partially shared systems of concepts among agents. These ontologies are defined for knowledge on object, action, and agents' abilities that are used in mediating given tasks (see Section 5).

## 4.1 Need for sharing concepts

The simplest way to accomplish a task with multiple agents is to break down the task and design subtasks each of which is executable for each agent. But this approach is not applicable where tasks are dynamically generated like environments where human and agents co-exist.

In order to do it more intelligently, agents should understand what parters are doing or requesting and so on. In other words, agents should have common communication capabilities to tell and understand intension. It means that they should share not only protocols and languages to communicate but also concepts used in their communication. The latter is called ontology which a system of concepts shared by agents to communicate to each other[6].

Ontology is defined and used mostly in information agents (For example see [10][2]). The primary concern in such studies is to model objects which agents should handle. Modeling objects is not sufficient to realize communication among real-world agents. Modeling space is also important because they should share space to cooperate each other. Modeling action is another important concept because they should understand what other agents do or request[1]. Therefore there are three ontologies, namely ontologies for object, space, and action.

## 4.2 Concept for object

The environments are usually fulfilled with various objects, and tasks are usually related to some of these objects. They should share concepts for objects, otherwise they cannot tell what they recognize or handle to other agents.

Difficulty lies that what they can understand are different because the way they can perceive objects is different. It depends on abilities for sensing, acting, and information processing.

The policy for making shared concepts is using abstraction levels to represent objects. We build a taxonomy of objects as hierarchy of *is-a* relations. It does not mean that all agents can understand all objects in the taxonomy. Most agents can only understand subsets of those objects because their recognition abilities are limited. For example, some agent can recognize a box but cannot recognize difference between a trash box and a parts-case, because it can only

---

[1] Another important concept is one for time. In this paper, time is not explicitly described but embedded as shared actions.

detect whether it is a box or not. It is sufficient for this agent to understand concept *box* and its higher concepts.

We provide current position, default position, color, weight for attributes which are common for all objects. Descriptions of attributes have also levels of abstraction.

## 4.3 Concept for space

The next important concept for cooperation is concept for space. Since agents are working in the physical world, they should understand space, that is, where they are, where they are moving for, where the target object exists, and so on. Especially it is important for agents to work together. According to sensing and processing abilities of agents, they can have different ways to represent space. For example, agents which move by programmed paths would represent space by paths and points to stop. Some agents can have absolute positions but some can have only relative positions. We provide the following two types of representation as shared space ontology.

1. **Representation with preposition**

   Relative position is described as combination of preposition and object which is represented as object ontology[7]. We provide seven prepositions, i.e., *at, on, in, in-front-of, behind, to-the-right-of,* and *to-the-left-of*. For example, a position in front of the rack agent is represented as `in-front-of` (`rack-agent`). Actual position is determined by agents who interpret representation.

2. **Representation with action**

   A relative position can be also represented as association to actions that can be performed at. For example, to describe space as "where you can look at the rack" or "where you can meet Agent X." is useful for agents who want to achieve these actions. Actual positions may be different according to agents that would take action, because ability of action that agent can do may be different. But no matter actual positions may differ, it is sufficient to understand positions where such actions can be done.

   We describe a position with combination of an action-related term and object(s). For example, `viewpoint(rack)` means a position where the agent can look at the rack, and `meetingpoint(agent1, agent2)` means where agent1 and agent2 can meet.

## 4.4 Concept for action

The last category for shared concepts is concept for action. Agents should understand what other agents are doing in order to cooperate with them. Like the other categories, concepts which an agent can understand are also different according to ability of the agent itself. It is obvious that concepts which are directly associated to its physical actions. But more abstract concepts can be shared among agents. Concepts associated to agents' physical actions should be related to more abstract concepts shared by them in order to understand each other.

Definition of concept for action consists of a name, attributes like subject and starting-point, and constraints among attributes. Constraints are represented as sharing of attribute values. Relation among concepts is decomposition relation, i.e., an action can have an sequence of action which can achieve the original action. Relation among decomposed actions are represented as constraints among attributes.

## 5 Ontology-based task mediation for cooperation

In this section, we discuss how to realize interaction among agents with different ontologies. We introduce mediators which can break down and translate tasks to a sequence of actions each of which some agent can understand and execute.

The function of mediators here is to bridge a gap between tasks provided by human and actions that can be done by real-world agents. Since tasks should be performed cooperatively by multiple agents in most cases, tasks should be decomposed into subtasks and distributed to agents. Ontologies have two roles in this process. Firstly, it is used to understand the given tasks. Since given tasks are what humans want agents to do, they are insufficient and incomplete for specifying actions of agents. Ontologies can supply necessary information on environments and agents to complete task descriptions. Secondly, it is used to distribute tasks to agents. As we mentioned in the previous section, each agent has its own ontology which is dependent on their physical and information ability. But shared ontologies can integrate these agent ontologies using abstraction. Tasks can be translated to a set of local tasks each of which is understandable by some agent by using multiple abstraction levels in ontologies.

We realized process of the mediation by the following four steps (see Figure 4). A task is described as an incomplete description of action. Incompleteness means that all properties should not be specified, i.e, some properties are specified, but others not. Unspecified properties will be fulfilled by mediators using the current state of the environment by consulting object ontology and object knowledge, e.g., where objects are now.

**Supplement of object attributes** If necessary attributes of objects are missing in a task description, the mediator can add these attributes using default values in object ontology.

**Assignment of agents** The mediator tries to assign an agent to perform the action to realize the task. It is done by consulting knowledge on agent abilities which is represented by object, space, and action ontologies.

**Action decomposition** The mediator decomposes the given action into actions each of which may be executable by some agents. Decomposition of action is done by consulting action ontology. Action decomposition and agent assignment are done simultaneously because action decomposition restricts

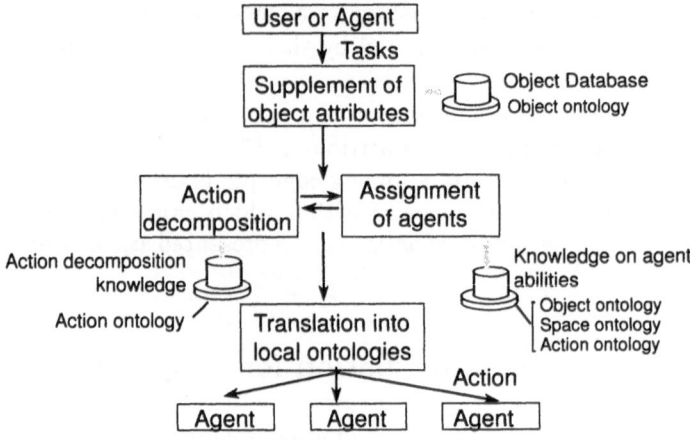

**Fig. 4.** Mediation flow

agent assignment and vice versa. If all actions are assigned to executable agents, the both steps are finished.

**Translation into local ontology** All information before this step is represented by the shared ontologies. Before sending out the decomposed actions to agents as a message, the mediator translates each message into one in the local ontology to the receiver agent.

The implemented mediator has two parts, i.e, planner and executor. Planner processes the above mediating steps, and executor binds agents participating the action sequence and control executing sequence of action (see Figure 5).

The above process describes how to deal with a single task. In the human-machine co-existing environment, there are multiple asynchronous tasks. In our approach, it is processed by cooperation among multiple mediators. The basic idea is that every emerged task invokes a mediator and then each mediator tries to gather and control necessary agents independently. Each mediator processes the given task by using state information of the environment and communication with other mediators if necessary (see Figure 5).

## 6 Various interaction between human and robots

We need natural ways for people to communicate and cooperate with machines or robots just as same as they do with other people, i.e., people interact with other people anywhere at anytime. In this section, we mainly focus on interaction between people and mobile robots.

The primitive way for human-robot interaction is interaction through special instruments. People can communicate with robots by using instruments like computers. Recent technologies for multimodal communication can provide various

155

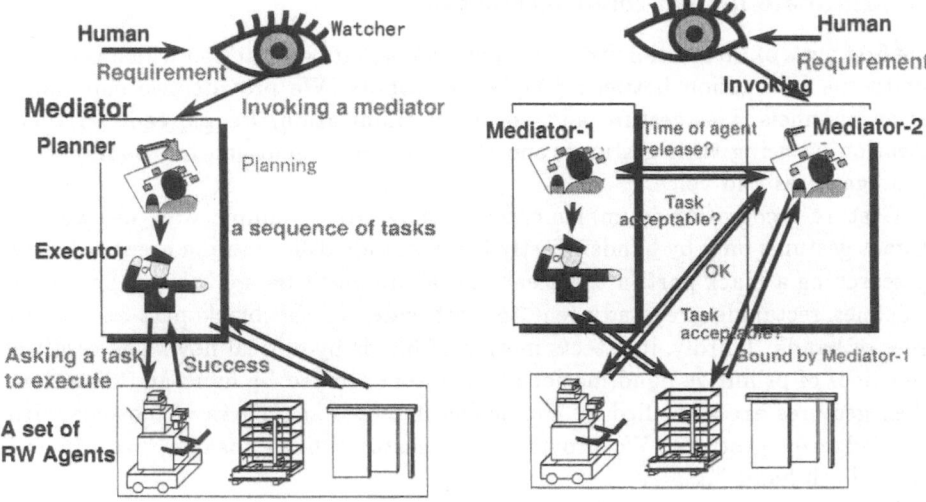

**Fig. 5.** Behavior of mediator(1)    **Fig. 6.** Behavior of mediator(2)

communication channels like voice and gestures(e.g., [3]). Interface agents (e.g., [8])can be used for their communication. But people could not communicate with robots directly, and they are bound to computer terminals.

Other way is direct interaction with people and robots. In addition to multimodal communication with computer, robots can use their bodies when they communicate to people. Although it is more restricted in expressive flexibility than *virtual* interface agents because of their mechanical structures, physical motion are more natural and acceptable for people. We call such direct interaction between robot and people *intimate interaction*.

The intimate interaction can involve people in multimodal direct interaction, but another problem arises. People and robots should be close to each other in order to establish such interaction. It is obstacle to realize ubiquitous interaction among people and robots. We need interaction between people and robots who are separate from each other. We call such interaction *loose interaction*.

Loose interaction absorbs the distance problem between people and robots, but interaction is still closed within participants of interaction. We sometimes need more robots (or even people) involved to accomplish interaction. For example, a robot is asked to bring a book by a person, but it has no capacity to bring books. It should ask another robot that can bring books and the person should interact another robot as a result. We call this type of interaction *cooperative interaction*. Cooperative interaction makes interaction extensive, i.e., interaction can be extended by introducing more robots and people as much as it needs. It can solve the problem of limitation of functions of each robot so that interaction should not be bound to functions of robots that people are interacting.

## 6.1   Intimate human-robot interaction

The first type of interaction we investigate is intimate interaction which is direct one-to-one interaction between people and robots. We provide two communication channels, i.e., gesture and vocal communication. People can tell their intention by using their gestures, and the real-world agent can tell its intention by its gestures and voice.

Gesture recognition is implemented in a relatively simple way, i.e, we can extract gestures only by hands. Firstly the agent identifies motion areas of hands by searching a black part in the scene and assuming it person's head. Secondly, it defines rectangle areas adjacent to both sides of the black part as motion areas of hands. Thirdly, it detects motion of hands by optical flow. The result is sequences of primitive hand motions which are specified by hand and direction. Then gestures are identified by comparing detected sequences of motions with knowledge on gestures. We provide some gestures like "shake", "wave", and "move both hands".

There needs another step to know meaning of such detected gestures, because meaning of gestures is dependent on situation of interaction. In our system, the real-world agent reacts to gestures according to predefined state transition network. Each state has actions that the real-world agent should take and some links to other states. Each link has conditions described with gestures of the person and its sensor modes. If one of conditions of link of the current state is satisfied, the current state is shifted to next state that is pointed by the link. Since a single gesture can be included in conditions of multiple links, multiple interpretation of gestures is possible. Figure 7 shows an example of intimate interaction.

Variety of actions that real-world agents can perform are classified into two. One is informative actions or gestures which cause no physical changes of the environment like "Yes", "No", and "Ununderstanding" using head motion, and "bye-bye" and "raise both hands" using hand motion. Voice generation is also included in possible informative actions of the real-world agent. The other is effective actions which cause physical changes of the environment like "grasp something" and "release something" using hand motion, and "move to somewhere" using driving units.

We currently provide some interaction modes like "take a box", "*janken*[2]", and "bye-bye". Some interaction is closed within the real-world agent and the person, but others not. If the latter case, the real-world agent should ask tasks to a mediator in order to involve other real-world agents. We will discuss this process as cooperative interaction later.

## 6.2   Loose human-robot interaction

Loose interaction is interaction between people and robots who are separated. Since robot may not see the person, the same method for intimate interaction is

---

[2] It is a children's game in which two or more person show one of three forms of hand to each other. The agent uses hand motions instead of forming hands.

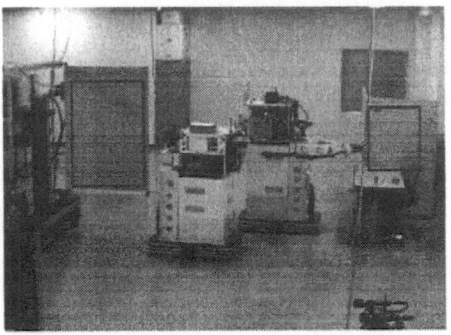

**Fig. 7.** An example of intimate interaction ("throw it out")

**Fig. 8.** Scene by camera for watcher (two boxes are where watcher is "watching".)

not applicable. We introduce an agent called "watcher" which *watches* a room to find what is happening in the room. It uses a camera to look over the room (see Figure 8) and communication to other agents.

If the watcher notices a request from someone to others, it composes a task description and passes to a mediator. Notification of requests comes by either recognition of camera scenes or communication from other agents. the watcher currently observes two areas, i.e., around a door and a desk (two boxes in Figure 8). An example of knowledge on task composition is shown in Figure 9. This definition tells "if it is found by camera that someone is waving, compose a task that Kappa1a should go to her/his position". As a result, the person who waves can tell her/his intention to the real-world agent even if it is not near her/him (see Figure 10). It is important that the watcher should not make direct order to real-world agents but tasks which can be scheduled by mediator. If the appointed agents are busy to process other tasks, the mediator can determine that the composed task may be postponed until the current task is finished, or be processed by other agents.

### 6.3 Cooperative human-robot interaction

Interaction should sometimes be extended to include agents needed to accomplish its purpose, i.e., interaction should be performed cooperatively by more than two agents. Suppose that a person is facing a robot that cannot take and carry objects and asking the robot to bring an object to her/him. The robot may try to do it by itself and finally finds it cannot, or simply refuse her/his request because it knows that it is impossible for it to do it. A better solution is that the robot should ask other robots that can take and carry objects to perform the recognized request. In this case, three agents, i.e., a person and two robots are necessary members to accomplish the interaction.

Cooperative human-robot interaction is realized here by mediators. If requests are detected by cameras, this process is done by watcher(see Figure 5).

```
(define Come_on
 (content
  ((behavior wave)
   (source    camera)
   (client    ?human)))
 (task
  ((subject camera)
   (come (subject kappa1a)
   (destination ?human)))))
```

**Fig. 9.** Knowledge on task composition

**Fig. 10.** An example of loose interaction (a camera behind the robot detected human request and told the robot to go)

Otherwise requesting agents themselves compose tasks and send them to the watcher. Then the watcher invokes a mediator and delegates the task to it.

Figure 11 shows how the example of cooperative interaction mentioned above can be solved in our system. In this example, two mediators are generated to solve a task with two mobile agents, a rack agent, and a person. In the example, the person asked a mobile agent to bring a manual on the rack. Unfortunately the mobile agent could not take objects on the rack. Then it asked to the watcher to solve the task. The mediator invoked by the watcher made a plan and executed it. On the other hand, watcher made another task to delegate another mediator because the mobile agent was obstacle for the first plan.

# 7  Related work

Most relevant studies are Robotic room[9] and Intelligent room[13][1]. Although they have similar goals but their methods are different in according to their application fields.

Robotic room is aiming intelligent environments for health care or hospitals. The key technology is to provide various sensoring devices and to detect human behaviors with them. It is different approach to ours in treatment of people in the system. People in their research are something for the system to observe, which is analogous to patients in hospitals.

Intelligent room project investigates various computational techniques to support people in meeting or discussion, for example, tracking person's movement and augmented reality that can impose computer-generated images to real images. People are here active and the system tries to help their activities, which is analogous to tools in offices.

On the other hand, the system and people are mutually understandable and cooperative in our system. Not only people can ask the system to help them,

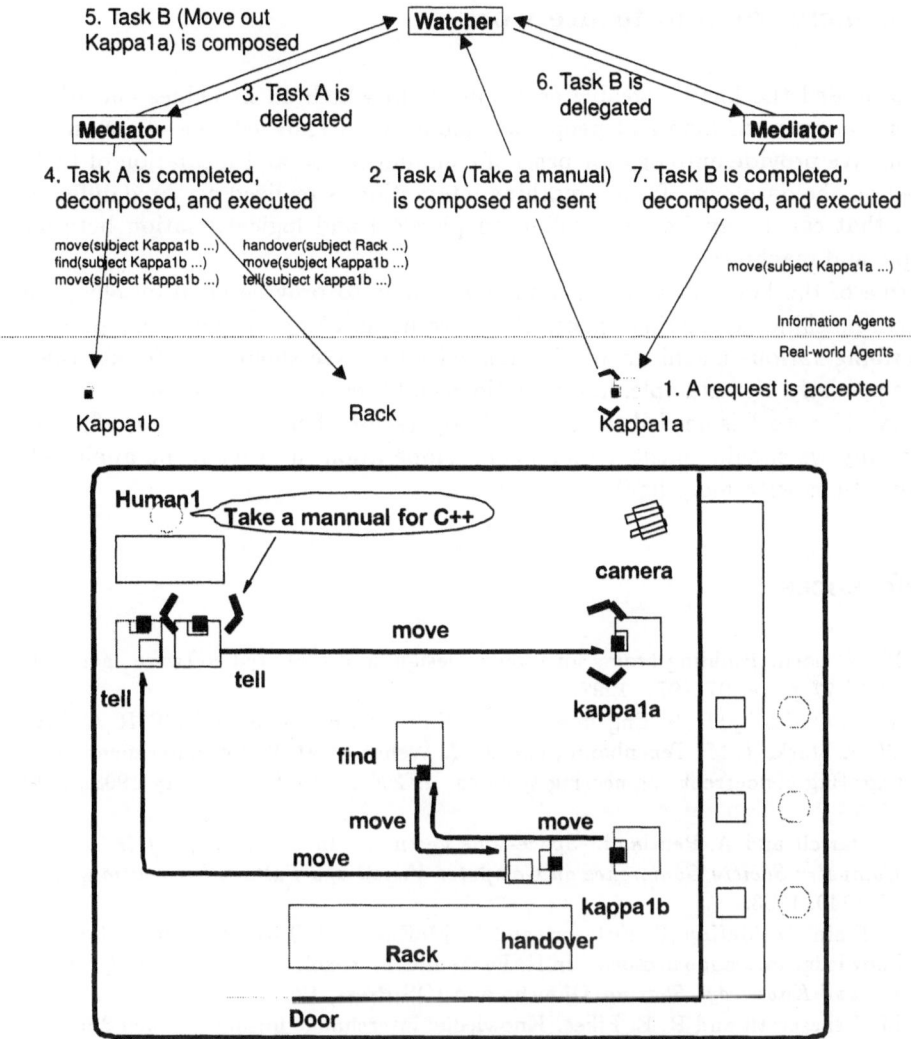

**Fig. 11.** An example of cooperative interaction

but the system may request people to help it when they are unable to perform the task asked by people. It is analogous to partners or secretaries in office.

It is interdisciplinary work so that there are much related work in artificial intelligence, robotics, and human interfaces. In particular there are interesting studies on human-robot interaction and cooperation of agents. Please see Takeda[12] in detail.

# 8 Conclusion and future work

We proposed the knowledgeable environment in which all machines and interaction between machine and people are modeled as knowledge-level communication. We provide ontology for basis of communication and mediation of tasks based on the ontology. Human-machine interaction is realized as three different ways that can be applied depending on physical and logical relation between people and machines.

One of the key issues in our approach is how to provide good ontology for human-machine co-existing space. The current ontology is naive and poor in describing various machines and human activities. We should investigate them more precisely. For example, human actions and their utilization of objects in ordinary office work is needed to analyze. Cooperation of agents is still insufficient, especially we should consider more tight cooperation in order to be applicable to situations with more limited resources.

# References

1. M. H. Coen. Building brains for rooms: Designing distributed software agents. In *IAAI-97*, pages 971–977, 1997.
2. M. R. Cutkosky, R. S. Engelmore, R. E. Fikes, M. R. Genesereth, T. R. Gruber, W. S. Mark, J. M. Tenenbaum, and J. C. Weber. PACT: An experiment in integrating concurrent engineering systems. *IEEE Computer*, January 1993:28–38, 1993.
3. T. Darrell and A. Pentland. Space-time gestures. In *Proceedings of IEEE 1993 Computer Society Conference on Computer Vision and Pattern Recognition*, pages 335–340, 1993.
4. T. Finin, D. McKay, R. Fritzson, and R. McEntire. KQML: An information and knowledge exchange protocol. In K. Fuchi and T. Yokoi, editors, *Knowledge Building and Knowledge Sharing*. Ohmsha and IOS Press, 1994.
5. M. Genesereth and R. E. Fikes. Knowledge interchange format, version 3.0 reference manual. Technical Report Technical Report Logic-92-1, Computer Science Department, Stanford University, June 1992.
6. T. R. Gruber. Toward principles for the design of ontologies used for knowledge sharing. Technical Report KSL 93-04, Knowledge Systems Laboratory, Stanford University, August 1993.
7. A. Herkovits. *Language and spatial cognition*. Cambridge University Press, 1986.
8. P. Maes and R. Kozierok. Learning interface agents. In *Proceedings of AAAI-93*, pages 459–465, 1993.
9. T. Sato, Y. Nishida, J. Ichikawa, Y. Hatamura, and H. Mizoguchi. Active understanding of human intention by a robot through monitoring of human behavior. In *Proceedings of the 1994 IEEE/RSJ International Conference on Intelligent Robots and Systems*, volume 1, pages 405–414, 1994.
10. H. Takeda, K. Iino, and T. Nishida. Agent organization and communication with multiple ontologies. *International Journal of Cooperative Information Systems*, 4(4):321–337, December 1995.

11. H. Takeda, K. Iwata, M. Takaai, A. Sawada, and T. Nishida. An ontology-based cooperative environment for real-world agents. In *Proceedings of Second International Conference on Multiagent Systems*, pages 353–360, 1996.

12. H. Takeda, N. Kobayashi, Y. Matsubara, and T. Nishida. Towards ubiquitous human-robot interaction. In *Working Notes for IJCAI-97 Workshop on Intelligent Multimodal Systems*, pages 1–8, 1997.

13. M. C. Torrance. Advances in human-computer interaction: The intelligent room. In *Working Notes of the CHI 95 Research Symposium*, Denvar, Colorado, 1995.

# Springer
# and the
# environment

At Springer we firmly believe that an international science publisher has a special obligation to the environment, and our corporate policies consistently reflect this conviction.
We also expect our business partners – paper mills, printers, packaging manufacturers, etc. – to commit themselves to using materials and production processes that do not harm the environment. The paper in this book is made from low- or no-chlorine pulp and is acid free, in conformance with international standards for paper permanency.

 Springer

# Lecture Notes in Artificial Intelligence (LNAI)

# Lecture Notes in Computer Science